Laboratory Manual

General, Organic, and Biological

Chemistry

Structures of Life

Karen C. Timberlake

Professor Emeritus, Los Angeles Valley College

Benjamin
Cummings

San Francisco • Boston • New York
Capetown • Hong Kong • London • Madrid • Mexico City
Montreal • Munich • Paris • Singapore • Sydney • Tokyo • Toronto

Acquisitions Editor: Maureen Kennedy
Project Editor: Claudia Herman
Managing Editor: Joan Marsh
Marketing Manager: Christy Lawrence
Manufacturing Coordinator: Vivian McDougal
Cover Design: Tony Asaro
Cover Photographs: John Bagley, Richard Tauber
Cover Illustration: Blakeley Kim

Some of the experiments in this lab book may be hazardous if materials are not handled properly or procedures are not followed correctly. Safety procedures of your college must be followed as directed by your instructor. Safety precautions must be utilized when you work with laboratory equipment, glassware, and chemicals.

ISBN 0-8053-2984-6

Benjamin
Cummings

11 12 13 14 15 CRS 08070605

www.aw.com/bc

Preface

Welcome to the *Laboratory Manual for General, Organic, and Biological Chemistry: Structures of Life*. In the process of writing lab manuals, I have developed experiments that illustrate each of the chemical principles we discuss from the first day of class. I have also taken care to make each experiment workable as well as providing critical thinking for the student. The goals of this laboratory manual address the following areas:

1. **Experiments relate to basic concepts of chemistry and health.** Experiments are designed to illustrate the chemical principles we discuss in our classes. They include experiments that relate to health and medicine, and often use common materials that are familiar to students.

2. **Experiments are flexible.** Each experiment includes a flexible group of sections, which allows instructors to select the sections to fit into their weekly laboratory schedule. Lab times and comments are given for each.

3. **Safety.** A detailed safety section in the preface includes a safety quiz. The aim here is to highlight the safety and equipment preparation on the first day of lab. In addition, each lab contains reminders of safety behavior. Students are reminded to wear goggles for every lab session. Some experiments are recommended as instructor demonstrations.

4. **Experiment format provides clear instructions and evaluation.** Each lab begins with a set of goals, a discussion of the topics, and examples of calculations. The report pages begin with pre-lab questions to prepare students for lab work. Students obtain data, draw graphs, make calculations, and write conclusions about their results. Each lab contains questions and problems that require the student to discuss the experiment, make additional calculations, and use critical thinking to apply concepts to real life.

5. **Stockroom preparation of chemicals.** Materials for each experiment are listed in the appendix with amounts given for 20 students working in pairs. Most lab sessions use standard lab equipment and chemicals that are readily available and inexpensive. In some cases students bring samples from home.

I hope that this laboratory manual will help you in your chemistry instruction and that students will find they learn chemistry by participating in the laboratory experience.

Karen C. Timberlake
Los Angeles Valley College
Valley Glen, CA 91401

To the Student

Here you are in a chemistry laboratory with your laboratory book in front of you. Perhaps you have already been assigned a laboratory drawer, full of glassware and equipment you may never have seen before. Looking around the laboratory, you may see bottles of chemical compounds, balances, burners, and other equipment that you are going to use. This may very well be your first experience with experimental procedures. At this point you may have some questions about what is expected of you. This laboratory manual is written with those considerations in mind.

The activities in this manual were written specifically to parallel the chemistry you are learning in the lecture portion of class. Many of the laboratory activities include materials that may be familiar to you, such as household products, diet drinks, cabbage juice, antacids, and aspirin. In this way, chemical topics are related to the real world and to your own science experience. Some of the labs teach basic skills; others encourage you to extend your scientific curiosity beyond the lab.

It is important to realize that the value of the laboratory experience depends on the time and effort you invest in it. Only then will you find that the laboratory can be a valuable learning experience and an integral part of the chemistry class. The laboratory gives you an opportunity to go beyond the lectures and words in your textbook and experience the scientific process from which conclusions and theories concerning chemical behavior are drawn. In some experiments, the concepts are correlated with health and biological concepts. Chemistry is not an inanimate science, but one that helps us to understand the behavior of living systems.

Using This Laboratory Manual

Each experiment begins with learning goals to give you an overview of the topics you will be studying in that experiment. Each experiment is correlated to concepts you are currently learning in your chemistry class. Your instructor will indicate which activities you are to do. At the beginning of each experiment, you will also find a list of the materials needed for each activity.

The experimental procedures are written to guide you through each laboratory activity. When you are ready to begin the lab, remove the report sheet at the end of the laboratory instructions. Place it next to the procedures for that section. Read and measure carefully, report your data, and follow instructions to complete the necessary calculations. You may also be asked to answer some or all of the follow-up questions and problems designed to test your understanding of the concepts from the experiment.

It is my hope that the laboratory experience will help illuminate the concepts you are learning in the classroom. The experimental process can help make chemistry a real and exciting part of your life and provide you with skills necessary for your future.

Contents

PART I: GENERAL CHEMISTRY

Contents

Contents

Working Safely in the Laboratory

The chemistry laboratory with its equipment, glassware, and chemicals has the potential for accidents. In order to avoid accidents, precautions must be taken by every student to ensure the safety of everyone working in the laboratory. By following the rules for handling chemicals safely and carrying out only the approved procedures, you will create a safe environment in the laboratory. After you have read the following sections, complete the safety quiz and the questions on laboratory equipment. Then sign and submit the commitment to lab safety.

A. Preparing for Laboratory Work

Pre-read Before you come to the laboratory, read the discussion of and directions for the experiment you will be doing. Make sure you know what the experiment is about before you start the actual work. If you have a question, ask your instructor to clarify the procedures.

Do assigned work only Do only the experiments that have been assigned by your instructor. No unauthorized experiments are to be carried out in the laboratory. Experiments are done at assigned times, unless you have an open lab situation. Your instructor must approve any change in procedure.

Do not work alone in a laboratory.

Safety awareness Learn the location and use of the emergency eyewash fountains, the emergency shower, fire blanket, fire extinguishers, and exits. Memorize their locations in the laboratory. Be aware of other students in the lab carrying chemicals to their desk or to a balance.

 APPROVED EYE PROTECTION IS REQUIRED AT ALL TIMES!

Safety goggles must be worn all the time you are in the lab The particular type depends on state law, which usually requires industrial-quality eye protection. Contact lenses may be worn in the lab if needed for therapeutic reasons, provided that **safety goggles** are worn over the contact lenses. Contact lenses without goggles are dangerous because splashed chemicals make them difficult to remove. If chemicals accumulate under a lens, permanent eye damage can result. If a chemical should splash into your eyes, flood the eyes with water at the eyewash fountain. Continue to rinse with water for at least 10 minutes.

Wear protective clothing Wear sensible clothing in the laboratory. Loose sleeves, shorts, or open-toed shoes can be dangerous. A lab coat is useful in protecting clothes and covering arms. Wear shoes that cover your feet to prevent glass cuts; wear long pants and long-sleeved shirts to protect skin. Long hair should be tied back so it does not fall into chemicals or a flame from a Bunsen burner.

No food or drink is allowed at any time in the laboratory Do not let your friends or children visit while you are working in the lab; have them wait outside.

Prepare your work area Before you begin a lab, clear the lab bench or work area of all your personal items, such as backpacks, books, sweaters, and coats. Find a storage place in the lab for them. All you will need is your laboratory manual, calculator, pen or pencil, text, and equipment from your lab drawer.

B. Handling Chemicals Safely

Check labels twice Be sure you take the correct chemical. ***DOUBLE-CHECK THE LABEL*** on the bottle before you remove a chemical from its container. For example, sodium sulfate (Na_2SO_4) could be mistaken for sodium sulfite (Na_2SO_3) if the label is not read carefully.

Use small amounts of chemicals Pour or transfer a chemical into a small, clean container (beaker, test tube, flask, etc.) available in your lab drawer. To avoid contamination of the chemical reagents, never insert droppers, pipets, or spatulas into the reagent bottles. Take only the quantity of chemical you need for the experiment. Do not keep a reagent bottle at your desk; *return* it to its proper location in the laboratory. Label the container. Many containers have etched sections on which you can write in pencil. If not, use tape or a marking pencil.

Do not return chemicals to the original containers To avoid contamination of chemicals, dispose of used chemicals according to your instructor's instructions. *Never return unused chemicals to reagent bottles.* Some liquids and water-soluble compounds may be washed down the sink with plenty of water, but check with your instructor first. Dispose of organic compounds in specially marked containers in the hoods.

Do not taste chemicals; smell a chemical cautiously Never use any equipment in the drawer such as a beaker to drink from. When required to note the odor of a chemical, first take a deep breath of fresh air and hold it while you use your hand to fan some vapors toward your nose and note the odor. Do not inhale the fumes directly. If a compound gives off an irritating vapor, use it in the fume hood to avoid exposure.

Do not shake laboratory thermometers Laboratory thermometers respond quickly to the temperature of their environment. Shaking a thermometer is unnecessary and can cause breakage.

Liquid spills Spills of water or liquids at your work area or floor should be cleaned up immediately. Small spills of liquid chemicals can be cleaned up with a paper towel. Large chemical spills must be treated with absorbing material such as cat litter. Place the contaminated material in a waste disposal bag and label it. If a liquid chemical is spilled on the skin, flood *immediately with water* for at least 10 minutes. Any clothing soaked with a chemical must be removed immediately because an absorbed chemical can continue to damage the skin.

Mercury spills The cleanup of mercury requires special attention. Mercury spills may occur from broken thermometers. Notify your instructor immediately of any mercury spills so that special methods can be used to clean up the mercury. Place any free mercury and mercury cleanup material in special containers for mercury only.

Laboratory accidents Always notify your instructor of any chemical spill or accident in the laboratory. Broken glass can be swept up with a brush and pan and placed in a specially labeled container for broken glass. Cuts are the most common injuries in a lab. If a cut should occur, wash, elevate, and apply pressure if necessary. Always inform your instructor of any laboratory accident.

Clean up Wash glassware as you work. Begin your cleanup 15 minutes before the end of the laboratory session. Return any borrowed equipment to the stockroom. Be sure that you always turn off the gas and water at your work area. Make sure you leave a clean desk. Check the balance you used. *Wash your hands before you leave the laboratory.*

C. Heating Chemicals Safely

Heat only heat-resistant glassware Only glassware marked Pyrex® or Kimax® can be heated; other glassware may shatter. To heat a substance in a test tube, use a test tube holder. Holding the test tube at an angle, move it continuously through the flame. Never point the open end of the test tube at anyone or look directly into it. A hot piece of iron or glass looks the same as it does at room temperature. Place a hot object on a tile or a wire screen to cool.

Flammable liquids Never heat a flammable liquid over an open flame. If heating is necessary, your instructor will indicate the use of a steam bath or a hot plate.

F. Laboratory Equipment

When experiments call for certain pieces of equipment, it is important that you know which item to select. Using the wrong piece of equipment can lead to errors in measurement or cause a procedure to be done incorrectly. For safety and proper results, you need to identify the laboratory items you will use in doing lab work.

Look for each of the items shown in the pictures of laboratory equipment on pages xvi–xvii. The quiz will help you learn some of the common laboratory equipment.

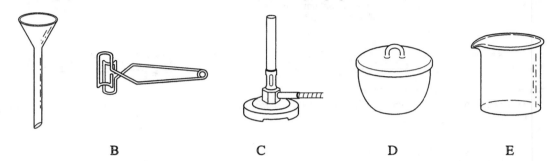

A   B   C   D   E

1.   Match each of the above pieces of equipment with its name:

_____ Bunsen burner   _____ test tube holder   _____ beaker

_____ crucible   _____ funnel

A   B   C   D   E

2.   Match each of the above pieces of equipment with its use in the laboratory:

_____ Used to hold liquids and to carry out reactions

_____ Used to pick up a crucible

_____ Used to support a beaker on an iron ring during heating

_____ Used to transfer small amounts of a solid substance

_____ Used to measure the volume of a liquid

A Visual Guide to Laboratory Equipment

Evaporating dish
Used to evaporate a solution to dryness

Crucible and cover
Used to heat small samples to high temperatures

Crucible tongs
Used to pick up a crucible

Watch glass
Used to cover a beaker or to hold a small amount of a substance

Stirring rod
Used to mix or combine two or more substances in a test tube or a beaker

Forceps
Used to pick up a small object or one that is hot

Pinch clamp
Used to close rubber tubing

Spatula
Used to transfer small amounts of a solid

File
Used to cut glass tubing

Thermometer
Used to measure the temperature of a substance

Pipet
Used to transfer a specific volume of liquid solution to a container

Buret
Used to deliver a measured amount of solution with a known concentration

Medicine dropper
Used to deliver drops of a solution or liquid

A Visual Guide to Laboratory Equipment

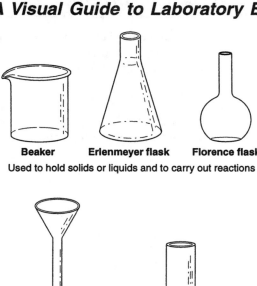

Beaker **Erlenmeyer flask** **Florence flask**
Used to hold solids or liquids and to carry out reactions

Graduated cylinder
Used to measure out accurate volumes of liquid

Wide-mouth bottle
Used to contain a gas or for reactions

Funnel
Used to transfer liquids to another container and when filtering out solids

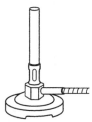

Shell vial
Used to hold small amounts of solid substances

Bunsen burner
Used to provide heat

Ring support stand with clay triangle
Used to support triangle or wire gauze during the heating process

Test tube
A small container for solutions and reactions

Test tube brush
Used to clean a test tube

Wire gauze
Used to support a beaker on an iron ring during heating

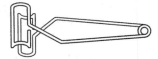

Test tube holder
Used to hold a test tube during heating or while still hot

Striker
Used to ignite the gas in a Bunsen burner

Test tube rack
Used to hold and store test tubes

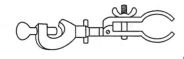

Clamp
Used to hold a buret, test tube, or flask to a ring stand

Heat-resistant tile
Used to set hot objects on during cooling

Graphing Experimental Data

When a group of experimental quantities is determined, a graph can be prepared that gives a pictorial representation of the data. After a data table is prepared, a series of steps is followed to construct a graph.

Preparing a Data Table

A data table is prepared from measurements. Suppose we measured the distance traveled in a given time by a bicycle rider. Table 1 is a data table prepared by listing the two variables, time and distance, that we measured.

Table 1 *Time and Distance Measurement*

Time (hr)	Distance (km)
1	5
3	14
4	20
6	30
7	33
8	40
9	46
10	50

Constructing the Graph

Draw vertical and horizontal axes Draw a vertical and a horizontal axis on the appropriate graph paper. The lines should be set in to leave a margin for numbers and labels, but the graph should cover most of the graph paper. Place a title at the top of the graph. The title should describe the quantities that will be placed on the axes.

Label each axis The label for each axis reflects the measurement listed in the data table. On our sample graph, the labels are time (hr) for the horizontal axis and distance (km) for the vertical axis.

Apply constant scales On each axis apply a scale of equal intervals that includes the full range of data points (low to high) you have in the data table. The intervals on a scale must be *equally spaced* and fit on the line you have drawn. Do not exceed the graph lines. Use intervals on each axis that are convenient counting units (2, 4, 6, 8, etc. or 5, 10, 15, etc.). The interval size on one axis does not need to match the size of the intervals on the other axis.

For our sample graph, we used a scale for a distance range of 0 km to 50 km. Each graph division represents 5 km. (You only have to number a few lines in order to interpret the scale. It gets too crowded with numbers if every line is marked.) Every two divisions on the time scale represent a time interval of 1 hour within the 10-hour time span of the bicycle ride. See Figure 1.

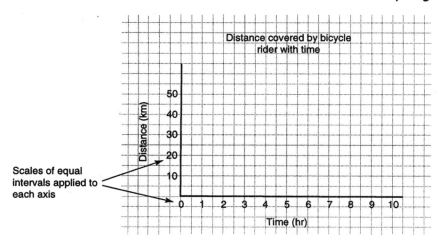

Figure 1 Marking equal intervals for distance and time on the axes

Plot the data points Plot the points for each pair of measurements on the data table. Follow a measurement on the vertical axis across until it meets a line that would be drawn from the corresponding measured value on the horizontal axis.

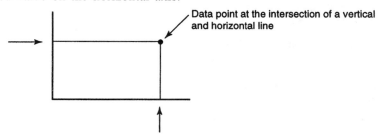

For example, at 4 hours, the rider has traveled 20 km. On the graph, find 20 km on the distance scale, and 4 hr on the time scale. Then follow the perpendicular lines to where they intersect. That is a point on the graph. Plotting each data pair will show the relationship between distance and time. A smooth line or curve is drawn that best fits the data points. However, some points may not fit on the line or curve you draw. That occurs when error is associated with the measurements or when the data are affected by other variables, such as terrain and energy level for the bicycle rider. See Figure 2.

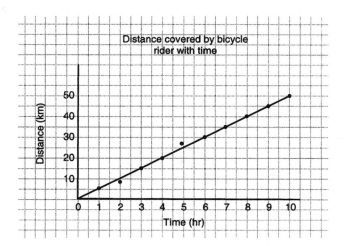

Figure 2 A completed graph with data points connected in a smooth line

Using the Laboratory Burner

Wear your goggles!

Materials: Bunsen burner, striker or matches

In the laboratory, substances are often heated with a Bunsen burner, shown in the figure below. The burner consists of a metal tube and base connected to a gas source. The flow of gas is controlled by adjusting the gas lever at the bench or by turning the wheel at the base of the burner. The amount of air that enters the burner is adjusted by twisting the tube to open or close the air vents. The gas and air mixture is ignited at the top of the tube using a match or a striker. Make sure the gas valves are tightly closed when you leave the laboratory after using the Bunsen burner.

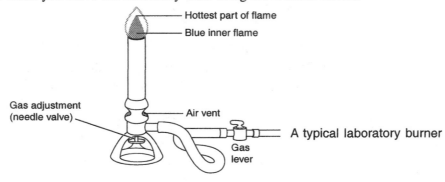

A typical laboratory burner

1. Before you light the burner, practice the following:
   a. Open and close the gas lever at the lab bench.
   b. Open and close the gas needle valve (wheel) at the base of the burner.
   c. Open and close the air vents.

   With the air vents closed, ignite the burner with a striker or match. Your instructor may demonstrate the use of the striker. Turn on the gas and hold the flame or spark at the top rim of the burner tube.

2. If the flame is yellow and sooty, the gas mixture does not have an adequate supply of oxygen. Open the air vents until the color of the flame changes to blue. Adjust the gas flow until you have a flame that is 6–8 cm high with two distinct parts, an inner cone and an outer flame.

3. The hottest part of a flame is at the tip of the inner blue flame. For the most effective heating, be sure that the tip of the inner flame is placed just under the substance you heat. *Remember what you heated: Hot metal and glass items do not look hot!*

Questions

Q.1 What is the color of the flame with the air vent closed? _____

Open? _____

How do you control the height of the flame? _____

_____

Q.2 What is the appearance of a flame that is used for heating? _____

_____

Q.3 Where is the hottest part of a flame? _____

Measurement and Significant Figures

Goals

- Identify metric units used in measurement such as gram, meter, centimeter, millimeter, and milliliter.
- Correctly read a meterstick, a balance, and a graduated cylinder.
- State the correct number of significant figures in a measurement.

Discussion

Scientists and allied health personnel carry out laboratory procedures, take measurements, and report results accurately and clearly. How well they do these things can mean life or death to a patient. The system of measurement used in science, hospitals, and clinics is the metric system. The metric system is a *decimal system* in which measurements of each type are related by factors of 10. You use a decimal system when you change U.S. money. For example, 1 dime is the same as 10 cents or one cent is 1/10 of a dime. A dime and a cent are related by a factor of 10.

The metric system has one standard unit for each type of measurement. For example, the metric unit of length is the meter, whereas the U.S. system of measurement uses many units of length such as inch, foot, yard, and mile. Most of the rest of the world uses the metric system only. The most common metric units are listed in Table 1.1.

Table 1.1   *Metric Units*

Measurement	Metric Unit	Symbol
Length	meter	m
Mass	gram	g
Volume	liter	L
Temperature	degrees Celsius; kelvins	°C; K
Time	second	s

A unit must always be included when reporting a measurement. For example, 5.0 m indicates a quantity of 5.0 meters. Without the unit, we would not know the system of measurement used to obtain the number 5.0. It could have been 5.0 feet, 5.0 kilometers, or 5.0 inches. Thus, a unit is required to complete the measurement reported.

For larger and smaller measurements, prefixes are attached in front of the standard unit. Some prefixes such as *kilo* are used for larger quantities; other prefixes such as *milli* are used for smaller quantities. The most common prefixes are listed in Table 1.2.

Table 1.2   *Some Prefixes in the Metric System*

Prefix	Symbol	Meaning
kilo	k	1000
deci	d	0.1 (1/10)
centi	c	0.01 (1/100)
milli	m	0.001 (1/1000)

Measured and Exact Numbers

When we measure the length, volume, or mass of an object, the numbers we report are called *measured numbers*. Suppose you got on a scale this morning and saw that you weighed 145 lb. The scale is a measuring tool and your weight is a measured number. Each time we use a measuring tool to determine a quantity, the result is a measured number.

    Exact numbers are obtained when we count objects. Suppose you counted 5 beakers in your laboratory drawer. The number 5 is an exact number. You did not use a measuring tool to obtain the number. Exact numbers are also found in the numbers that define a relationship between two metric units or between two U.S. units. For example, the numbers in the following definitions are exact: 1 meter is equal to 100 cm; 1 foot has 12 inches. See Sample Problem 1.1.

Sample Problem 1.1

Describe each of the following as a measured or exact number:

a. 14 inches   b. 14 pencils   c. 60 minutes in 1 hour   d. 7.5 kg

Solution:

a. measured   b. exact   c. exact (definition)   d. measured

Significant Figures in Measurements

In measured numbers, all the reported figures are called *significant figures*. The first significant figure is the first nonzero digit. The last significant figure is always the estimated digit. Zeros between other digits or at the end of a decimal number are counted as significant figures. However, leading zeros are *not significant;* they are placeholders. Zeros are *not significant* in large numbers with no decimal points; they are placeholders needed to express the magnitude of the number.

    When a number is written in scientific notation, all the figures in the coefficient are significant. Examples of counting significant figures in measured numbers are in Table 1.3 and Sample Problem 1.2.

Table 1.3 *Examples of Counting Significant Figures*

Measurement	Number of Significant Figures	Reason
455.2 cm	4	All nonzero digits are significant.
0.80 m	2	A following zero in a decimal number is significant.
50.2 L	3	A zero between nonzero digits is significant.
0.0005 lb	1	Leading zeros are not significant.
25,000 ft	2	Placeholder zeros are not significant.

Sample Problem 1.2

State the number of significant figures in each of the following measured numbers:

a. 0.00580 m   b. 132.08 g

Solution:

a. Three significant figures. The zeros after the decimal point are placeholder zeros, but the zero following nonzero digits is *significant.*
b. Five significant figures. The zero between nonzero digits is significant.

When you use a meterstick or read the volume in a graduated cylinder, the measurement must be reported as precisely as possible. The number of *significant figures* you can report depends on the lines marked on the measuring tool you use. For example, on a 50-mL graduated cylinder, the small lines represent a 1-mL volume. If the liquid level is between 21 mL and 22 mL, you know you can report 21 mL for certain. However, you can add one more digit (*the last digit*) to your reported value by estimating between the 1-mL lines. For example, if the volume level were halfway between the 21-mL and 22-mL lines, you would report the volume as 21.5 mL. If the volume level *is exactly* on the 21-mL line, you indicate this precision by adding a *significant zero* to give a measured volume of 21.0 mL.

A. Measuring Length

The standard unit of length in the metric system is the *meter* (*m*). Using an appropriate prefix, you can indicate a length that is greater or less than a meter as listed in Table 1.4. Kilometers are used in most countries for measuring the distance between two cities, whereas centimeters or millimeters are used for small lengths.

Table 1.4 *Some Metric Units Used to Measure Length*

Length	Symbol	Meaning
1 kilometer	km	1000 meters (m)
1 decimeter	dm	0.1 m (1/10 m)
1 centimeter	cm	0.01 m (1/100 m)
1 millimeter	mm	0.001 m (1/1000 m)

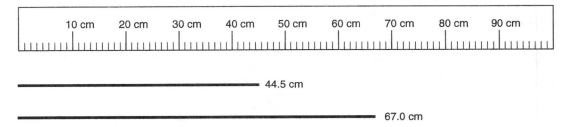

Figure 1.1 A meterstick divided into centimeters (cm)

A *meterstick* is divided into 100 cm as seen in Figure 1.1. The smallest lines are centimeters. That means that each measurement you make can be certain to the centimeter. The final digit in a precise measurement is obtained by estimating. For example, the shorter line in Figure 1.1 reaches the 44-cm mark and is about halfway to 45 cm. We report its length as 44.5 cm. The last digit (0.5) is the estimated digit. If the line appears to end at a centimeter mark, then the estimated digit is 0.0 cm. The longer line in Figure 1.1 appears to end right at the 67-cm line, which is indicated by reporting its length as 67.0 cm.

B. Measuring Volume

The volume of a substance measures the space it occupies. In the metric system, the unit for volume is the *liter* (*L*). Prefixes are used to express smaller volumes such as deciliters (dL) or milliliters (mL). One cubic centimeter (cm^3 or cc) is equal to 1 mL. The terms are used interchangeably. See Table 1.5.

Table 1.5 *Some Metric Units Used to Measure Volume*

Unit of Volume	Symbol	Meaning
1 kiloliter	kL	1000 liters (L)
1 deciliter	dL	0.1 L (1/10 L)
1 milliliter	mL	0.001 L (1/1000 L)

In the laboratory, the volume of a liquid is measured in a graduated cylinder (Figure 1.2). Set the cylinder on a level surface and bring your eyes even with the liquid level. Notice that the water level is not a straight line but curves downward in the center. This curve, called a *meniscus,* is read at its lowest point (center) to obtain the correct volume measurement for the liquid. In this graduated cylinder, the volume of the liquid is 42.0 mL.

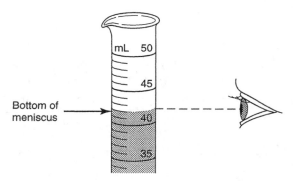

Figure 1.2   Reading a volume of 42.0 mL in a graduated cylinder

On large cylinders, the lines may represent volumes of 2 mL, 5 mL, or 10 mL. On a 250-mL cylinder, the marked lines usually represent 5 mL. On a 1000-mL cylinder, each line may be 10 mL. Then your precision on a measurement will be to the milliliter or mL.

C.   Measuring Mass

The *mass* of an object indicates the amount of matter present in that object. The *weight* of an object is a measure of the attraction that Earth has for that object. Because this attraction is proportional to the mass of the object, we will use the terms *mass* and *weight* interchangeably.

In the metric system, the unit of mass is the *gram (g)*. A larger unit, the *kilogram (kg),* is used in measuring a patient's weight in a hospital. A smaller unit of mass, the *milligram (mg),* is often used in the laboratory. See Table 1.6.

Table 1.6   *Some Metric Units Used to Measure Mass*

Mass	Symbol	Meaning
kilogram	kg	1000 g
gram	g	1000 mg
milligram	mg	1/1000 g (0.001 g)

Lab Information

Time:	2 hr

Comments:	Tear out the report sheets and place them beside the experimental procedures as you work.
	Determine the markings on each measuring tool before you measure.
	Record all the possible numbers for a measurement including an estimated digit.
	Write a unit of measurement after each measured number.

Related Topics: Significant figures, measured and exact numbers, metric prefixes

Experimental Procedures

A. Measuring Length

Materials: Meterstick, string

A.1 Observe the marked lines on a meterstick. Identify the lines that represent centimeters and millimeters. Determine how you will estimate between the smallest lines.

A.2 Use the meterstick to make the length measurements (cm) indicated on the report sheet. String may be used to determine the distance around your wrist. Include the estimated digit in each measurement.

A.3 Indicate the estimated digit and the number of significant figures in each measurement. See Sample Problem 1.3.

Sample Problem 1.3
What is the estimated digit in each of the following measured masses?
a. beaker 42.18 g   b. pencil 11.6 g

Solution:
a. hundredths place (0.08 g)   b. tenths place (0.6 g)

A.4 Measure the length of the line on the report sheet. List the measurements of the same line obtained by other students in the lab.

B. Measuring Volume

Materials: Display of graduated cylinders with liquids, 50-mL, 100-mL, 250-mL, and 500-mL (or larger) graduated cylinders, test tube, solid object

B.1 **Volume of a liquid** Determine the volumes of the liquids in a display of graduated cylinders. Be as precise as you can. For example, each line marked on a 50-mL graduated cylinder measures 1 mL. By estimating the volume *between* the 1-mL markings, you can report a volume to a tenth (0.1) of a milliliter. Indicate the estimated digit and the number of significant figures in each measurement.

B.2 **Volume of a test tube** Fill a small test tube to the rim. Carefully pour the water into a small graduated cylinder. State the volume represented by the smallest marked lines on the cylinder. Record the volume of the water and state the estimated digit. Fill the test tube again and pour the water into a medium-sized graduated cylinder. Record. Repeat this process using a large graduated cylinder. Record.

B.3 **Volume of a solid by volume displacement** When an object is submerged in water, it displaces its own volume of water, causing the water level to rise. The volume of the object is the difference in the water level before and after the object is submerged. See Figure 1.3.

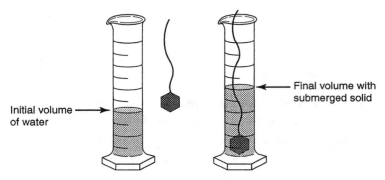

Figure 1.3 Using volume displacement to determine the volume of a solid

Obtain a graduated cylinder that will hold the solid. Place water in the graduated cylinder until it is about half full. Carefully record the volume of water. Tie a piece of thread around a heavy solid object. Slowly submerge the solid under the water. Record the new volume of the water. Calculate the volume (mL) displaced by the solid.

C. Measuring Mass

Materials: Balance, objects to weigh (beaker, rubber stopper, evaporating dish), unknown mass

C.1 After your instructor shows you how to use a laboratory balance, determine the mass of the listed objects from your lab drawer. If you are using a triple beam balance, be sure that all of your recorded measurements include an estimated digit.

C.2 Now that you have used the balance several times, obtain an object of unknown mass from your instructor. Record the code number and determine its mass. Record. Check your result with the instructor.

Conversion Factors in Calculations

Goals

- Round off a calculated answer to the correct number of significant figures.
- Determine the area of a rectangle and the volume of a solid by direct measurement.
- Determine metric and metric-to-U.S.-unit equalities and corresponding conversion factors.
- Use conversion factors in calculations to convert units of length, volume, and mass.
- Convert a Celsius temperature to Fahrenheit and Kelvin.

Discussion

As you begin to perform laboratory experiments, you will make measurements, collect data, and carry out calculations. When you use measured numbers in calculations, the answers that you report must reflect the precision of the original measurements. Thus it is often necessary to adjust the results you see on the calculator display. Every time you use your calculator, you will need to assess the mathematical operations, count significant figures, and round off calculator results.

A. Rounding Off

Usually there are fewer significant figures in the measured numbers used in a calculation than there are digits that appear in a calculator display. Therefore, we adjust the calculator result by rounding off. If the first number to be dropped is *less than 5,* it and all following numbers are dropped. If the first number to be dropped is *5 or greater,* the numbers are dropped and the value of the last *retained* digit is increased by 1. When you round a large number, the correct magnitude is retained by replacing the dropped digits with *placeholder zeros.* See Sample Problem 2.1. When a whole number appears in the calculator display, significant zeros may be added.

Sample Problem 2.1
Round off each of the following calculator displays to report an answer with three significant figures and another answer with two significant figures:
a. 75.6243   b. 0.528392   c. 387,600   d. 4

Solution:	**Three Significant Figures**	**Two Significant Figures**
a.	75.6	76
b.	0.528	0.53
c.	388,000	390,000
d.	4.00	4.0

B. Significant Figures in Calculations

When you carry out mathematical operations, the answer you report depends on the number of significant figures in the data you used.

Multiplication/division When you multiply or divide numbers, report the answer with the same number of significant figures as the measured number with the *fewest* significant figures (the *least precise*). See Sample Problem 2.2.

Sample Problem 2.2

Solve: $\dfrac{0.025 \times 4.62}{3.44} =$

Solution:

On the calculator, the steps are

$0.025 \times 4.62 \div 3.44$ = 0.033575581   *calculator display*

                    = 0.034   *final answer rounded to two significant figures*

Addition/subtraction When you add or subtract numbers, the reported answer has the same number of decimal places as the measured number with the *fewest* decimal places. See Sample Problem 2.3.

Sample Problem 2.3

Add: 2.11 + 104.056 + 0.1205

Solution: 2.11   *two decimal places*

            104.056

            $\underline{0.1205}$

            106.2865 *calculator display*

            106.29   *final answer rounded to two decimal places*

C.   Conversion Factors for Length

Metric factors If a quantity is expressed in two different metric units, a *metric equality* can be stated. For example, the length of 1 meter is the same as 100 cm, which gives the *equality* 1 m = 100 cm. When the values in the equality are written as a fraction, the ratio is called a *conversion factor.* Two fractions are always possible, and both are conversion factors for the equality.

Equality	*Conversion Factors*
1 m = 100 cm	$\dfrac{100 \text{ cm}}{1 \text{ m}}$ and $\dfrac{1 \text{ m}}{100 \text{ cm}}$

Metric–U.S. system factors When a quantity measured in a metric unit is compared to its measured quantity in a U.S. unit, a *metric–U.S.* conversion factor can be written. For example, 1 inch is the same length as 2.54 cm, as seen in Figure 2.1.

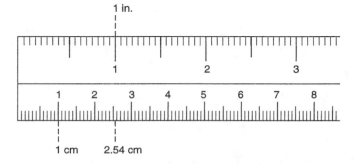

Figure 2.1 Comparing centimeters and inches

For the length of 1 inch, two conversion factors can be written.

Equality	*Conversion Factors*
1 in. = 2.54 cm	$\dfrac{2.54 \text{ cm}}{1 \text{ in.}}$ and $\dfrac{1 \text{ in.}}{2.54 \text{ cm}}$

D.   Conversion Factors for Volume

In the metric system, equalities for volume can be written along with their corresponding conversion factors. A useful metric–U.S. equality is the relationship of 1 quart equaling 946 mL.

Equality	*Conversion Factors*
1 L = 1000 mL	$\dfrac{1000\ mL}{1\ L}$ and $\dfrac{1\ L}{1000\ mL}$
1 qt = 946 mL	$\dfrac{946\ mL}{1\ qt}$ and $\dfrac{1\ qt}{946\ mL}$

E.   Conversion Factors for Mass

In the metric system, equalities for mass can be written along with their corresponding conversion factors. A useful metric–U.S. equality is the relationship of 454 g equaling 1 pound.

Equality	*Conversion Factors*
1 kg = 1000 g	$\dfrac{1000\ g}{1\ kg}$ and $\dfrac{1\ kg}{1000\ g}$
1 lb = 454 g	$\dfrac{454\ g}{1\ lb}$ and $\dfrac{1\ lb}{454\ g}$

F.   Percent by Mass

A percent (%) by mass gives the number of grams of each component in 100 grams of the mixture. It is calculated by dividing the mass of each component by the mass of the mixture and multiplying by 100.

$$\frac{\text{Mass (g) of component 1}}{\text{Mass (g) of the mixture}} \quad \times \quad 100 \quad = \quad \%\ \text{of component 1}$$

$$\frac{\text{Mass (g) of component 2}}{\text{Mass (g) of the mixture}} \quad \times \quad 100 \quad = \quad \%\ \text{of component 2}$$

G.   Converting Temperature

Temperature measures the intensity of heat in a substance. A substance with little heat feels cold. Where the heat intensity is great, a substance feels hot. The temperature of our bodies is an indication of the heat produced. An infection may cause body temperature to deviate from normal. On the Celsius scale, water freezes at 0°C; on the Fahrenheit scale, water freezes at 32°F. A Celsius temperature is converted to its corresponding Fahrenheit temperature by using the following equation:

$$T_F = 1.8\ (T_C) + 32$$

When the Fahrenheit temperature is known, the Celsius temperature is determined by rearranging the equation. Be sure you subtract 32 from the T_F, then divide by 1.8.

$$T_C = \frac{(T_F - 32)}{1.8}$$

A Celsius temperature can be converted to a Kelvin temperature by using the following equation:

$$T_K = T_C + 273$$

Lab Information

Time: 3 hr
Comments: Tear out the report sheets and place them beside the procedures.
                  Determine what the smallest lines of measurement are on each measuring tool you use.
                  Include an estimated digit for each measurement.
                  Round off the calculator answers to the correct number of significant figures.
Related Topics: Conversion factors, significant figures in mathematical operations, calculator use

Experimental Procedures

A. Rounding Off

Materials: Meterstick, solid

A.1 **Rounding** A student has rounded off some numbers. Determine whether the rounding was done correctly. If it is incorrect, state what the student needs to do and write the correctly rounded number.

A.2 **Area** Determine the length (cm) and width (cm) of the sides of the rectangle on the report sheet. Obtain a second set of measurements from another student. Record. Calculate the area (cm^2) of the rectangle using your measurements and this formula: Area = L × W. Obtain the area calculated by the other student. Compare the calculated areas from both sets of measurements.

A.3 **Volume of a solid by direct measurement** Obtain a solid object that has a regular shape, such as a cube, rectangular solid, or cylinder. Record its shape. Use a meterstick to determine the dimensions of the solid in centimeters (cm). Use the appropriate formula from the following list to calculate the volume in cm^3.

Shape	Dimensions to Measure	Formulas for Volume
Cube	Length (L)	$V = L^3$
Rectangular solid	Length (L), width (W), height (H)	$V = L \times W \times H$
Cylinder	Diameter (D), height (H)	$V = \dfrac{\pi D^2 H}{4} = \dfrac{3.14 D^2 H}{4}$

B. Significant Figures in Calculations

B.1 Solve the multiplication and division problems. Report your answers with the correct number of significant figures.

B.2 Solve the addition and subtraction problems. Report your answers with the correct number of significant figures.

C. Conversion Factors for Length

Materials: Meterstick

C.1 **Metric factors** Observe the markings for millimeters on a meterstick. Write an equality that states the number of millimeters in 1 meter. Write two metric conversion factors for the relationship. Observe the number of millimeters in a centimeter. Write equality and corresponding conversion factors for the relationship between centimeters and millimeters.

C.2 **Metric–U.S. system factors** Measure the length of the dark line on the report sheet in centimeters and in inches. Convert any fraction to a decimal number. Divide the number of centimeters by the number of inches to give a relationship. Round off correctly for your reported answer. This is your *experimental* value for the number of centimeters in 1 inch.

C.3 **Your metric height** Record your height in inches. Or use a yardstick to measure. Using the appropriate conversion factors, *calculate* your height in centimeters and meters. Show your setup for each calculation.

$$\text{Height (in.)} \times \frac{2.54 \text{ cm}}{1 \text{ in.}} = \text{your height (cm)}$$

$$\text{Height (cm)} \times \frac{1 \text{ m}}{100 \text{ cm}} = \text{your height (m)}$$

D. Conversion Factors for Volume

Materials: 1-L graduated cylinder, 1-quart measure (or two 1-pint measures)

D.1 Observe the markings on a 1-liter graduated cylinder. Write an equality that states the number of milliliters in 1 liter. Write two conversion factors for the equality.

D.2 Using a 1-pint or 1-quart measure, transfer 1 quart of water to a 1-liter graduated cylinder. Record the number of milliliters in 1 quart. Write the equality that states the number of milliliters in a quart. Write two conversion factors for the equality.

E. Conversion Factors for Mass

Materials: Commercial product with mass (weight) of contents given on label

E.1 **Grams and pounds** Labels on commercial products list the amount of the contents in both metric and U.S. units. Obtain a commercial product. Record the mass (weight) of the contents stated on the product label. *Do not weigh.* If the weight is given in ounces, convert it to pounds (1 lb = 16 oz).

$$_____ \text{ oz} \times \frac{1 \text{ lb}}{16 \text{ oz}} = _____ \text{ lb}$$

Divide the grams of the product by its weight in pounds. (Be sure to use the correct number of significant figures.) This is your value for grams in one pound (g/lb).

E.2 **Pounds and kilograms** State the mass on the label in kilograms. If necessary, convert the number of grams to kilograms.

$$_____ \text{ g} \times \frac{1 \text{ kg}}{1000 \text{ g}} = _____ \text{ kg}$$

Divide the number of pounds by the number of kilograms. Report the ratio as lb/kg. (Be sure to use the correct number of significant figures.)

F. Percent by Mass

> **Materials**: 100- or 250-mL beaker, graduated cylinder
> Sucrose (sugar), water

F.1 Weigh a 100-mL or 250-mL beaker or tare the beaker. Record.

> ***Taring a container on an electronic balance:*** The mass of a container on an electronic balance can be set to 0 by pressing the tare bar. As a substance is added to the container, the mass shown on the readout is for the substance only. (When a container is tared, it is not necessary to subtract the mass of the beaker.)

F.2 Add about 5 g of sugar. Record the mass.

F.3 With the beaker and sugar still on the balance, add 15–20 mL of water to the sugar in the beaker. Record the mass of the sugar–water mixture.

F.4 Calculate the % sugar and % water by mass in the sugar–water mixture.

G. Converting Temperature

Materials: Thermometer (°C), a 150- or 250-mL beaker, ice, and rock salt

G.1 Observe the markings on a thermometer. Indicate the lowest and highest temperatures that can be read using that thermometer. ***Caution: Never shake down a laboratory thermometer. Shaking a laboratory thermometer can cause breakage and serious accidents.***

G.2 To measure the temperature of a liquid, place the bulb of the thermometer in the center of the solution. Keep it *immersed* while you read the temperature scale. When the temperature becomes constant, record the temperature (°C). On most thermometers, you can estimate the tenths of a degree (0.1°C). A set of beakers with the following contents may be set up in the lab; otherwise fill the beakers as instructed. Determine the temperature of each of the following:

a. Room temperature: Place the thermometer on the lab bench.

b. Tap water: Fill a 250-mL beaker about 1/3 full of water.

c. Ice-water mixture: Add enough ice to the water in part b to double (approximately) the volume. Allow 5 minutes for the temperature to change.

d. A salted ice mixture: Add rock salt to the ice-water mixture in part c. Stir and allow a few minutes for the temperature to change.

G.3 Convert the Celsius temperatures to corresponding temperatures on the Fahrenheit and Kelvin scales.

Report Sheet - Lab 2

Date _____ January 22, '06 _____   Name LAUREN-NICOLE D. PASCUAL

Section CHEM-1301-AA _____   Team _____

Instructor _____   _____

Pre-Lab Study Questions

1. What are the rules for rounding off numbers?

2. How do you determine how many significant figures to keep in an answer obtained by multiplying or dividing?

   WHEN MULTIPLYING OR DIVIDING, WRITE THE ANSWER WITH THE SAME NUMBER OF SIG. FIGURES AS THE MEASURED # W/ THE FEWEST SIG. FIGURES

3. How is the number of digits determined for an answer obtained by adding or subtracting?

   WHEN ADDING OR SUBTRACTING, THE ANSWER MUST HAVE THE SAME # OF DECIMAL PLACES AS THE MEASURED # WITH THE FEWEST DECIMAL PLACES.

4. Is a body temperature of 39.4°C a normal temperature or does it indicate a fever?

   $T_F = 1.8(39.4) + 32$ → INDICATES A FEVER!
   $= 102.92°F$

5. What is an equality and how is it used to write a conversion factor?

   – AN EQUALITY IS A QUANTITY EXPRESSED IN TWO DIFFERENT METRIC UNITS.
   – WHEN THE VALUES IN THE EQUALITY ARE WRITTEN AS A FRACTION, THE RATIO IS CALLED A CONVERSION FACTOR.

A. Rounding Off

A.1 Rounding A student rounded off the following calculator displays to three significant figures. Indicate if the rounded number is correct. If it is incorrect, round off the display value properly.

Calculator Display	Student's Rounded Value	Correct (yes/no)	Corrected (if needed)
24.4704	24.5	YES	
143.63212	144	YES	
532,800	530	NO	
0.00858345	0.009	YES	
8	8.00	YES	

Report Sheet - Lab 2

A.2 Area

	Your measurements	Another student's measurements
Length =	_____	_____
Width =	_____	_____
Area =	_____	_____

(Show calculations.)

Why could two students obtain different values for the calculated areas of the same rectangle?

A.3 Volume of a solid by direct measurement

Shape of solid _____

Formula for volume of solid _____

height _____   **length** _____

width _____   **diameter** *(if cylinder)* _____

Volume of the solid _____
(Show calculations of volume including the units.)

Report Sheet - Lab 2

B. Significant Figures in Calculations

B.1 Perform the following multiplication and division calculations. Give a final answer with the correct number of significant figures:

4.5×0.28 _____

$0.1184 \times 8.00 \times 0.0345$ _____

$\dfrac{(42.4)(15.6)}{1.265}$ _____

$\dfrac{(35.56)(1.45)}{(4.8)(0.56)}$ _____

B.2 Perform the following addition and subtraction calculations. Give a final answer with the correct number of significant figures.

13.45 mL + 0.4552 mL _____

145.5 m + 86.58 m + 1045 m _____

1315 + 200 + 1100 _____

245.625 g – 80.2 g _____

4.62 cm – 0.885 cm _____

Questions and Problems

Q.1 What is the combined mass in grams of objects that have masses of 0.2000 kg, 80.0 g, and 524 mg?

Q.2 A beaker has a mass of 225.08 g. When a liquid is added to the beaker, the combined mass is 238.254 g. What is the mass in grams of the liquid?

Report Sheet - Lab 2

C.   Conversion Factors for Length

C.1   Metric factors

    Equality: 1 m   = _____ mm

Conversion factors: $\dfrac{\boxed{}\ \text{m}}{\boxed{}\ \text{mm}}$ and $\dfrac{\boxed{}\ \text{mm}}{\boxed{}\ \text{m}}$

Equality: 1 cm   = _____ mm

Conversion factors: $\dfrac{\boxed{}\ \text{cm}}{\boxed{}\ \text{mm}}$ and $\dfrac{\boxed{}\ \text{mm}}{\boxed{}\ \text{cm}}$

C.2   Metric–U.S. system factors

_____

Line length (*measured*)   _____ in.

                               _____ cm

$\dfrac{\boxed{}\ \text{cm}}{\boxed{}\ \text{in.}}$ = $\dfrac{\boxed{}\ \text{cm}}{1\ \text{in.}}$ (*Experimental ratio*)

How close is your *experimental ratio* to the standard conversion factor of 2.54 cm/in.?

Report Sheet - Lab 2

C.3 **Your metric height**

Height (inches) _____

Height in centimeters *(calculated)*

_____ in. × $\dfrac{\underline{\hspace{2cm}} \text{cm}}{1 \text{ in.}}$ = _____ cm

What is your height in meters? _____ m
(Show your calculations here.)

Questions and Problems *(Show complete setups.)*

Q.3 A pencil is 16.2 cm long. What is its length in millimeters (mm)?

Q.4 A roll of tape measures 45.5 inches. What is the length of the tape in meters?

D. **Conversion Factors for Volume**

D.1 Equality: 1 L = _____ mL

Conversion factors: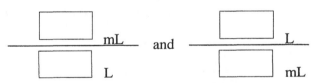

D.2 Volume (mL) of 1 quart of water: _____ mL

Number of milliliters in 1 quart: _____ mL/qt *(experimental)*

Equality: 1 qt = $\dfrac{\underline{\hspace{2cm}} \text{mL}}{}$

Conversion factors: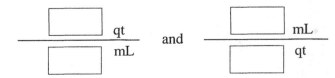

Report Sheet - Lab 2

Questions and Problems *(Show complete setups.)*

Q.5   A patient received 825 mL of fluid in one day. What is that volume in liters?

Q.6   How many liters of plasma are present in 8.5 pints? (1 qt = 2 pt)

E.   Conversion Factors for Mass

E.1   **Grams and pounds**

Name of commercial product   _____

Mass in grams stated on label   _____

Weight in lb or oz given on label   _____

Weight in lb
(*Convert oz to lb if needed.*)   _____

$$\frac{\text{Number of grams}}{\text{Number of lb}} = \frac{\boxed{}\ \text{g}}{\boxed{}\ \text{lb}} = \frac{\boxed{}\ \text{g}}{1\ \text{lb}}$$

How does your experimental factor compare to the standard value of 454 g/lb?

E.2   **Pounds and kilograms**

Mass in kilograms (from label)   _____

Weight in lb   _____

$$\frac{\text{Number of lb}}{\text{Number of kg}} = \frac{\boxed{}\ \text{lb}}{\boxed{}\ \text{kg}} = \frac{\boxed{}\ \text{lb}}{1\ \text{kg}}$$

How does your *experimental factor* compare to the standard value of 2.20 lb/kg?

Questions and Problems

Q.7   An infant has a mass of 3.40 kg. What is the weight of the infant in pounds?

Report Sheet - Lab 2

F.   Percent by Mass

F.1   Mass of the beaker *(0 if tared)*   _____

F.2   Mass of the sugar + the beaker   _____

F.3   Mass of the sugar–water mixture + beaker   _____

F.4   **Calculations:**

What is the mass of sugar?   _____

What is the mass of the sugar–water mixture?   _____

What is the mass of the water added?   _____

What is the % sugar (by mass)?
(Show calculations.)   _____

What is the % water (by mass)?
(Show calculations.)   _____

Questions and Problems *(Show complete setups)*

Q.8   A sugar–water mixture contains 45.8 g of sugar and 108.5 g of water. What is the percent by mass of sugar and the percent by mass of water in the solution?

G. Converting Temperature

G.1 Temperature scale(s) on the laboratory thermometer _____

Lowest temperature _____ Highest temperature _____

G.2   °C   (G.3) °F   K

    a. Room temperature   _____   _____   _____

    b. Tap water   _____   _____   _____

    c. Ice-water mixture   _____   _____   _____

    d. Salt ice-water mixture   _____   _____   _____

Questions and Problems

Q.9 Write an equation for each of the following temperature conversions:

    a. °C to °F

    b. °F to °C

    c. °C to K

Q.10 A recipe calls for a baking temperature of 205°C. What temperature in °F should be set on the oven?

Density and Specific Gravity

Goals

- Calculate the density of a substance from measurements of its mass and volume.
- Calculate the specific gravity of a liquid from its density.
- Determine the specific gravity of a liquid using a hydrometer.

Discussion

A. Density of a Solid

To determine the density of a substance, you need to measure both its mass and its volume. You have carried out both of these procedures in previous labs. From the mass and volume, the density is calculated. If the mass is measured in grams and the volume in milliliters, the density will have the units of g/mL.

$$\text{Density of a substance} = \frac{\text{Mass of substance}}{\text{Volume of substance}} = \frac{\text{g of substance}}{\text{mL of substance}}$$

B. Density of a Liquid

To determine the density of a liquid, you need the mass and volume of the liquid. The mass of a liquid is determined by weighing. The mass of a container is obtained and then a certain volume of liquid is added and the combined mass determined. Subtracting the mass of the container gives the mass of the liquid. From the mass and volume, the density is calculated.

$$\text{Density of liquid} = \frac{\text{Mass (g) of liquid}}{\text{Volume (mL) of liquid}}$$

C. Specific Gravity

The specific gravity of a liquid is a comparison of the density of that liquid with the density of water, which is 1.00 g/mL (4°C).

$$\text{Specific gravity (sp gr)} = \frac{\text{Density of liquid (g/mL)}}{\text{Density of water (1.00 g/mL)}}$$

Specific gravity is a number with no units; the units of density (g/mL) have canceled out. This is one of the few measurements in chemistry written without any units.

Using a hydrometer The specific gravity of a fluid is determined by using a hydrometer. Small hydrometers (urinometers) are used in the hospital to determine the specific gravity of urine. Another type of hydrometer is used to measure the specific gravity of the fluid in your car battery. A hydrometer placed in a liquid is spun slowly to keep it from sticking to the sides of the container. The scale on the hydrometer is read at the lowest (center) point of the meniscus of the fluid. Read the specific gravity on the hydrometer to 0.001. See Figure 3.1.

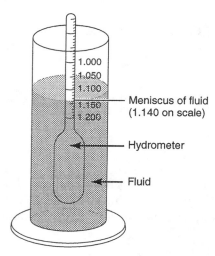

Figure 3.1 Measuring specific gravity using a hydrometer

D. Graphing Mass and Volume

When a group of experimental quantities are determined, a graph can be prepared that gives a pictorial representation of the data. After a data table is prepared, a series of steps are followed to construct a graph. See Graphing Experiment Data page xviii.

Lab Information

Time: 3 hr

Comments: Tear out the report sheets and place them beside the procedures.
             Round off the calculator answers to the correct number of significant figures.
             Dispose of liquids properly as directed by your instructor.

Related Topics: Mass, volume, prefixes, significant figures, density, specific gravity

Experimental Procedures      GOGGLES REQUIRED!

A. Density of a Solid

Materials: Metal object, string or thread, graduated cylinder

A.1 **Mass of the solid** Obtain a solid metal object. Determine its mass and record.

A.2 **Volume of the solid by displacement** Obtain a graduated cylinder that is large enough to hold the solid metal object. Add water until the cylinder is about half full. Read the water level carefully and record. If the solid object is heavy, lower it into the water by attaching a string or thread. While the solid object is submerged in the water, record the final water level. Calculate the volume of the solid.

Volume of solid = Final water level – initial water level

A.3 **Calculating the density of the solid** Calculate the density (g/mL) of the solid by dividing its mass (g) by its volume (mL). Be sure to determine the correct number of significant figures in your calculated density value.

$$\text{Density of solid} = \frac{\text{Mass (g) of solid}}{\text{Volume (mL) of solid}}$$

A.4 If your instructor indicates that the solid is made of one of the substances in Table 3.1, use the density you calculated in A.3 to identify the metal from the known values for density.

Table 3.1 *Density Values of Some Metals*

Substance	Density (g/mL)
Aluminum	2.7
Brass	8.4
Copper	8.9
Iron	7.9
Lead	11.3
Nickel	8.9
Tin	7.3
Zinc	7.1

B. Density of a Liquid

Materials: 50-mL graduated cylinder, two liquid samples, 100-mL or 250-mL beaker

B.1 **Volume of liquid** Place about 20 mL of water in a 50-mL graduated cylinder. Record. *(Do not use the markings on beakers to measure volume; they are not precise.)*

B.2 **Mass of liquid** The mass of a liquid is found by weighing by difference. First, determine the mass of a small, dry beaker. Pour the liquid into the beaker, and reweigh. Record the combined mass. *Be sure to write down all the figures in the measurements.* Calculate the mass of the liquid.

> *Taring a container on an electronic balance:* The mass of a container on an electronic balance can be set to 0 by pressing the *tare* bar. As a substance is added to the container, the mass shown on the readout is for the *substance* only. (When a container is *tared,* it is not necessary to subtract the mass of the beaker.)

B.3 **Density of liquid** Calculate the density of the liquid by dividing its mass (g) by the volume (mL) of the liquid.

$$\text{Density of liquid} = \frac{\text{Mass (g) of liquid}}{\text{Volume (mL) of liquid}}$$

Repeat the same procedure for another liquid provided in the laboratory.

C. Specific Gravity

Materials: Water, liquids used in part B in graduated cylinders with hydrometers

C.1 Calculate the specific gravity (sp gr) of each liquid you used in B. Divide its density by the standard density of water (1.00 g/mL).

$$\text{Specific gravity} = \frac{\text{Density of a substance (g/mL)}}{\text{Density of water (1.00 g/mL)}}$$

C.2 Read the hydrometer set in a graduated cylinder containing the same liquid you used in the density section. Record. Some hydrometers use the European decimal point, which is a comma. The value 1,000 on a European scale is read as 1.000. Record specific gravity as a decimal number.

D. Graphing Mass and Volume

Materials: Metal pieces such as aluminum, copper, or zinc or pennies (pre-1980 or post-1980) 50-mL graduated cylinder

In this graphing activity, we will show the relationship between the mass and volume of a substance. The volume and mass of five different samples of the same substance will be measured. After the data for the samples are collected, the mass and volume of each sample will be used to prepare a graph. The density (g/mL) will be visually represented on a graph.

D.1 Place about 20–25 mL of water in a 50-mL graduated cylinder. Carefully record the initial volume of water.

D.2 Place the cylinder and water on a top-loading balance and determine their mass. Record. Use the same balance to complete the experiment.

D.3 Add two or three pieces of metal or pennies. Record the new level of the water. Record the new mass. Subtract the initial volume of water from this water level to determine the volume of the metal pieces or the pennies. If you did not tare the cylinder and water originally, determine the mass of the metal pieces or pennies by subtracting the initial mass from this combined mass. Add some more pieces of metal or some more pennies to the cylinder. Each time record the resulting water level and the new mass. Repeat this process for a total of five sets of data.

D.4 Prepare a graph by plotting the mass (g) of the metal pieces or pennies on the vertical axis and the volume (mL) of the metal pieces or pennies on the horizontal axis. Use a ruler to draw a line through the points you have plotted. If some of the points fall off the line, run the line between them so you have as many points above the line as you have below the line. Draw a smooth line through the points.

D.5 The slope of the line on the graph represents the density of the metal. Mark two places on the line. Divide the difference between the two mass values by the difference of the two values for volume.

$$\frac{\text{Mass (2)} - \text{Mass (1)}}{\text{Volume (2)} - \text{Volume (1)}} \quad = \quad \frac{\text{g}}{\text{mL}} \quad = \quad \text{density of metal or pennies}$$

Report Sheet - Lab 3

Date _____   Name _____

Section _____   Team _____

Instructor _____   _____

Pre-Lab Study Questions

1. What property of oil makes it float on water?

2. Why would heating the gas in an air balloon make the balloon rise?

3. What is the difference between density and specific gravity?

4. How does a graph help us interpret scientific data?

A.   Density of a Solid

A.1 **Mass of the solid**   _____

A.2 **Volume of the solid by displacement**

    Initial water level (mL)   _____

    Final water level with solid (mL)   _____

    Volume of solid (mL)   _____

A.3 **Calculating the density of the solid**   _____ g/mL
    (Show calculations.)

A.4 Type of metal   _____

Questions and Problems *(Show complete setups.)*

Q.1 An object made of aluminum has a mass of 8.37 g. When it was placed in a graduated cylinder containing 20.0 mL of water, the water level rose to 23.1 mL. Calculate the density and specific gravity of the object.

Report Sheet - Lab 3

B. Density of a Liquid

		Liquid 1	Liquid 2
B.1	**Volume of liquid**		
	Type of liquid	_____	_____
	Volume (mL)	_____	_____
B.2	**Mass of liquid**		
	Mass of beaker	_____	_____
	Mass of beaker + liquid	_____	_____
	Mass of liquid	_____	_____
B.3	**Density of liquid**		
	Density	_____	_____
	(Show calculations for density.)		

C. Specific Gravity

C.1	Specific gravity	_____	_____
	(Calculated using B.3)		
C.2	Specific gravity	_____	_____
	(Hydrometer reading)		

How does the *calculated* specific gravity compare to the hydrometer reading for each liquid?

Questions and Problems *(Show complete setups.)*

Q.2 What is the mass of a solution that has a density of 0.775 g/mL and a volume of 50.0 mL?

Q.3 What is the volume of a solution that has a specific gravity of 1.2 and a mass of 185 g?

Report Sheet - Lab 3

D. Graphing Mass and Volume

D.1 Type of metal _____

Initial volume of water (mL) _____

D.2 Initial mass of cylinder + water (g) _____

D.3 **Mass of Metal Pieces   Final Volume   Total Volume of Metal (mL)**

_____ g   _____ mL   _____ mL

_____ g   _____ mL   _____ mL

_____ g   _____ mL   _____ mL

_____ g   _____ mL   _____ mL

_____ g   _____ mL   _____ mL

D.4 **Graph**

Mass of Metal Objects vs. Volume

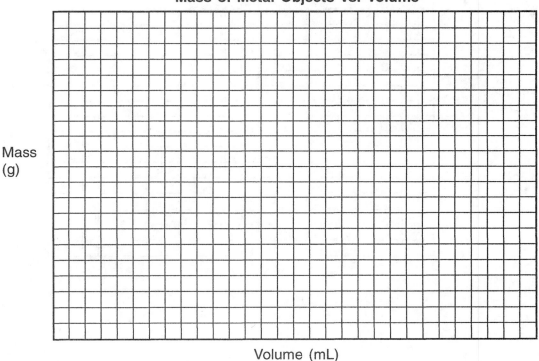

Mass
(g)

Volume (mL)

D.5 $\dfrac{\text{Mass (2)} - \text{Mass (1)}}{\text{Volume (2)} - \text{Volume (1)}}$ =   _____ –
                                                                    –

=   _____ g/mL

Report Sheet - Lab 3

Questions and Problems

Q.4 An IV pump delivers the following volume of saline solution over 4 hours.

Volume (mL)	Time (hours)
0	0
50	1.0
100	2.0
125	2.5
150	3.0
200	4.0

Prepare a graph to represent the data above.

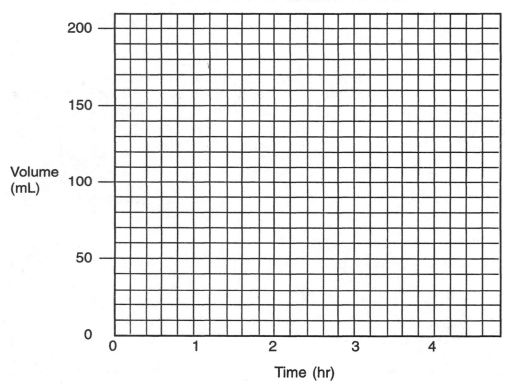

Volume of IV Solution vs. Time

Atomic Structure

Goals

- Write the correct symbols or names of some elements.
- Describe some physical properties of the elements you observe.
- Categorize an element as a metal or nonmetal from its physical properties.
- Given the complete symbol of an atom, determine its mass number, atomic number, and the number of protons, neutrons, and electrons.

Discussion

Primary substances, called elements, build all the materials about you. Some look similar, but others look unlike anything else. In this experiment, you will describe the physical properties of elements in a laboratory display and determine the location of elements on a blank periodic table.

A. Physical Properties of Elements

Metals are elements that are usually shiny or have a metallic luster. They are usually good conductors of heat and electricity, ductile (can be drawn into a wire), and malleable (can be molded into a shape). Some metals such as sodium or calcium may have a white coating of oxide formed by reacting with oxygen in the air. If these are cut, you can see the fresh shiny metal underneath. In contrast, nonmetals are not good conductors of heat and electricity, are brittle (not ductile), and appear dull, not shiny.

B. Periodic Table

The periodic table, shown on the inside front cover of this lab manual and your textbook, contains information about each of the elements. On the table, the horizontal rows are *periods,* and the vertical columns are *groups.* Each group contains elements that have similar physical and chemical properties. The groups are numbered across the top of the chart. Elements in Group 1 are the *alkali metals,* elements in Group 2 are the *alkaline earths,* and Group 7 contains the *halogens.* Group 8 contains the *noble gases,* which are elements that are not very reactive compared to other elements. A dark zigzag line that looks like a staircase separates the *metals* on the left side from the *nonmetals* on the right side.

C. Subatomic Particles

There are different kinds of atoms for each of the elements. Atoms are made up of smaller bits of matter called *subatomic particles. Protons* are positively charged particles, *electrons* are negatively charged, and *neutrons* are neutral (no charge). In an atom, the protons and neutrons are tightly packed in the tiny center called the *nucleus.* Most of the atom is empty space, which contains fast-moving electrons. Electrons are so small that their mass is considered to be negligible compared to the mass of the proton or neutron. The *atomic number* is equal to the number of protons. The *mass number* of an atom is the number of protons and neutrons.

> atomic number = number of protons (p^+)
> mass number = sum of the number of protons and neutrons ($p^+ + n^0$)

D. Isotopes

Isotopes are atoms of the same element that differ in the number of neutrons. This means that isotopes of an element have the same number of protons, but different mass numbers. The following example represents the symbol of a sulfur isotope that has 16 protons and 18 neutrons.

Complete Symbol of an Isotope	**Meaning**
mass number (p^+ and n^0) → **34**	This atom has 16 protons and 18 neutrons.
symbol of element → **S**	The element is sulfur.
atomic number (p^+) → **16**	The atom has 16 protons.

Lab Information

Time: 2 hr

Comments: Obtain a periodic table as a reference.
Tear out the report sheets and place them beside the procedures.
Carefully observe the physical properties of the elements in the display.

Related Topics: Names and symbols of the elements, periodic table, atoms, subatomic particles, isotopes, electrons and protons

Experimental Procedures

A. Physical Properties of Elements

Materials: A display of elements

Observe the elements in the laboratory display of elements. In the report sheet, write the symbol and atomic number for each element listed. Describe some physical properties such as color and luster. From your observations, identify each element as a metal (M) or a nonmetal (NM).

B. Periodic Table

Materials: Periodic table, colored pencils, display of elements

B.1 On the incomplete periodic table provided in the report sheet, write the atomic numbers and symbols of the elements you observed in part A. Write the group number at the top of each column of the representative (Groups 1–8) elements. Write the period numbers for each of the horizontal rows shown. Using different colors, shade in the columns that contain the alkali metals, alkaline earths, halogens, and noble gases. With another color, shade in the transition elements. Draw a heavy line to separate the metals and nonmetals.

B.2 *Without looking* at the display of elements, use the periodic table to decide whether the elements listed on the report sheet would be metals or nonmetals; shiny or dull. *After* you complete your predictions, observe those same elements in the display to see if you predicted correctly.

C. Subatomic Particles

For each of the neutral atoms described in the table, write the atomic number, mass number, and number of protons, neutrons, and electrons.

D. Isotopes

Complete the information for each of the isotopes of calcium: the complete nuclear symbol and the number of protons, neutrons, and electrons.

Report Sheet - Lab 4

Date _____   Name _____

Section _____   Team _____

Instructor _____   _____

1. Describe the periodic table.

2. Where are the alkali metals and the halogens located on the periodic table?

3. On the following list of elements, circle the symbols of the transition elements and underline the symbols of the halogens:

    Mg   Cu   Br   Ag   Ni   Cl   Fe   F

4. Complete the list of names of elements and symbols:

Name of Element	Symbol	Name of Element	Symbol
Potassium			Na
Sulfur			P
Nitrogen			Fe
Magnesium			Cl
Copper			Ag

Report Sheet - Lab 4

A. Physical Properties of Elements

Element	Symbol	Atomic Number	Color	Luster	Metal/Nonmetal
				Physical Properties	
Aluminum	_____	_____	_____	_____	_____
Carbon	_____	_____	_____	_____	_____
Copper	_____	_____	_____	_____	_____
Iron	_____	_____	_____	_____	_____
Magnesium	_____	_____	_____	_____	_____
Nickel	_____	_____	_____	_____	_____
Nitrogen	_____	_____	_____	_____	_____
Oxygen	_____	_____	_____	_____	_____
Phosphorus	_____	_____	_____	_____	_____
Silicon	_____	_____	_____	_____	_____
Silver	_____	_____	_____	_____	_____
Sulfur	_____	_____	_____	_____	_____
Tin	_____	_____	_____	_____	_____
Zinc	_____	_____	_____	_____	_____

Report Sheet - Lab 4

B. Periodic Table

B.1

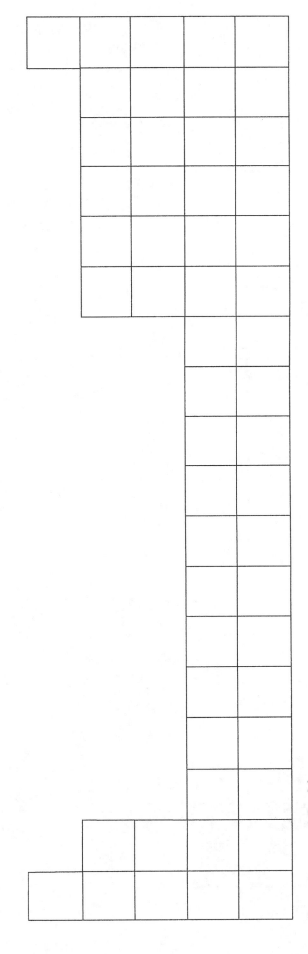

Questions and Problems

Q.1 From their positions on the periodic table, categorize the following elements as metals (M) or nonmetals (NM).

Na _____ S _____ Cu _____ F _____ Fe _____ C _____ Ca _____

Q.2 Give the name of each of the following elements:

a. Noble gas in Period 2 _____ b. Halogen in Period 2 _____

c. Alkali metal in Period 3 _____ d. Halogen in Period 3 _____

e. Alkali metal in Period 4 _____

Report Sheet - Lab 4

B.2

Element	Metal/Nonmetal	Prediction: Shiny or Dull	Correct? Yes/No
Chromium			
Gold			
Lead			
Cadmium			
Silicon			

C. Subatomic Particles

Element	Atomic Number	Mass Number	Protons	Neutrons	Electrons
Iron				30	
		27			13
			19	20	
Bromine		80			
Gold		197			
			53	74	

D. Isotopes

Nuclear Symbol	Protons	Neutrons	Electrons
$^{40}_{20}$Ca			
	20	22	
$^{43}_{20}$Ca			
		24	20
$^{46}_{20}$Ca			

Report Sheet - Lab 4

Questions and Problems

Q.3   A neutral atom has a mass number of 80 and has 45 neutrons. Write its complete symbol.

Q.4   An atom has two more protons and two more electrons than the atom in question 3. What is its complete symbol?

Electron Configuration and Periodic Properties

Goals

- Describe the color of a flame produced by an element.
- Use the color of a flame to identify an element.
- Write the electron configuration for an element.
- Draw a graph of atomic diameter against atomic number.
- Interpret the trends in atomic radii within a family and a period.

Discussion

A. Flame Tests

The chemistry of an element strongly depends on the arrangement of the electrons. The energy levels for electrons of atoms of the *first 20 elements* have the following number of electrons.

Electron Arrangement for Elements 1–20
Level 1 ($2e^-$)   Level 2 ($8e^-$)   Level 3 ($8e^-$)   Level 4 ($2e^-$)

When electrons absorb specific amounts of energy, they can attain higher energy levels. In order to return to the lower, more stable energy levels, electrons release energy. If the energy released is the same amount as the energy that makes up visible light, the element produces a color.

When heated, many of the elements in Groups 1A and 2A produce colorful flames. Each element produces a characteristic color. When the light from one of these flames passes through a glass prism or crystal, a series of color lines appears. The spaces between lines appear dark. Such a series of lines, known as a *spectrum,* is used to identify elements in water, food, the sun, stars, and on other planets.

B. Electron Configuration

In an electron configuration, electrons are arranged by subshells starting with the lowest energy. The number of electrons in each subshell is written as a superscript. The electron arrangement of an element is related to its position in the periodic table. The electron configuration can be written by following the subshell blocks across the periodic table starting with period 1. The *s* block is formed by Groups 1A and 2A. The *p* block includes the elements in Groups 3A to 8A. The period number gives the particular energy level of each p subshell beginning with $2p$. Examples are as follows:

Li $1s^2 2s^1$	**O** $1s^2 2s^2 2p^4$	**Ne** $1s^2 2s^2 2p^6$
Na $1s^2 2s^2 2p^6 3s^1$	**S** $1s^2 2s^2 2p^6 3s^2 3p^4$	**Ar** $1s^2 2s^2 2p^6 3s^2 3p^6$

On the periodic table, the 4s block fills next

K $1s^2 2s^2 2p^6 3s^2 3p^6 4s^1$ **Ca** $1s^2 2s^2 2p^6 3s^2 3p^6 4s^2$

The d block begins with atomic number 21 and includes ten transition metals. The energy level of each d block is one less than its period number.

Sc $1s^2 2s^2 2p^6 3s^2 3p^6 4s^2 3d^1$ **Fe** $1s^2 2s^2 2p^6 3s^2 3p^6 4s^2 3d^6$ **Zn** $1s^2 2s^2 2p^6 3s^2 3p^6 4s^2 3d^{10}$

Ga $1s^2 2s^2 2p^6 3s^2 3p^6 4s^2 3d^{10} 4p^1$ **As** $1s^2 2s^2 2p^6 3s^2 3p^6 4s^2 3d^{10} 4p^3$

Br $1s^2 2s^2 2p^6 3s^2 3p^6 4s^2 3d^{10} 4p^5$ **Kr** $1s^2 2s^2 2p^6 3s^2 3p^6 4s^2 3d^{10} 4p^6$

The *f* **block**, which has a maximum of 14 electrons, follows the 6s block. The energy level of each *f* block is two less than the corresponding period number.

C. Graphing a Periodic Property: Atomic Radius

Since the 1800s scientists have recognized that chemical and physical properties of certain groups of elements tend to be similar. A Russian scientist, Dmitri Mendeleev, found that the chemical properties of elements tended to recur when the elements were arranged in order of increasing atomic mass. This repetition of similar characteristics is called periodic behavior. He used this periodic pattern to predict the characteristics of elements that were not yet discovered. Later, H. G. Moseley established that the similarities in properties were associated with the atomic number.

In the electron arrangement of an element, the electrons in the highest or outermost energy level are called the valence electrons. The valence electrons determine the chemical properties of the elements. If the elements are grouped according to the number of valence electrons, their chemical and physical properties are similar. The similarities of behavior occur periodically as the number of valence electrons is repeated.

In this exercise, you will graph the relationship between the atomic radius of an atom and its atomic number. Such a graph will show a repeating or periodic trend. Observe the graph in Figure 5.1, which was obtained by plotting the average temperature of the seasons. The graph shows that a cycle of high and low temperatures repeats each year. Such a tendency is known as a periodic property. There are three cycles on this particular graph, one full cycle occurring every year. When such cycles are known, the average temperatures for the next year could be predicted.

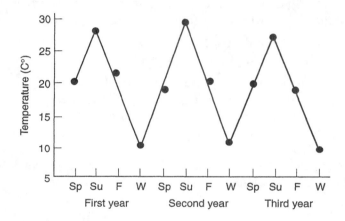

Figure 5.1 A graph of average seasonal temperatures.

Lab Information

Time: $1^1/_2$ hr

Comments: Obtain a periodic table or use the inside cover of your textbook.
                Tear out the Lab 5 report sheets and place them beside the procedures.
                In neutral atoms, the number of electrons is equal to the number of protons.

Related Topics: Electrons and protons, energy levels, and electron arrangement

Experimental Procedures

 **GOGGLES REQUIRED!**

A. Flame Tests

Materials: Bunsen burner, spot plate, flame-test (nichrome) wire, cork, 100-mL beaker, 1 M HCl, 0.1 M solutions (dropper bottles): $CaCl_2$, KCl, $BaCl_2$, $SrCl_2$, $CuCl_2$, NaCl, and unknown solutions

Obtain a spot plate, flame-test wire, and cork stopper. Bend one end of the flame-test wire into a small loop and secure the other end in a cork stopper. Pour a small amount of 1 M HCl into a 100-mL beaker. Rinse the spot plate in distilled water. Place 6–8 drops of each test solution in separate indentations of the spot plate. Label the spot plate diagram in the laboratory report to match the solutions. Be careful not to mix the different solutions.

CAUTION 1 M HCl is corrosive! Be careful when you use it. Wash off any HCl spills on the skin with tap water for 10 minutes.

Adjust the flame of a Bunsen burner until it is nearly colorless. Clean the test wire by dipping the loop in the HCl in the beaker and placing it in the flame of the Bunsen burner. If you see a strong color in the flame while heating the wire, dip it in the HCl again. Repeat until the color is gone.

Observing Flame Colors

Dip the cleaned wire in one of the solutions on the spot plate. Make sure that a thin film of the solution adheres to the loop. See Figure 5.2. Move the loop of the wire into the lower portion of the flame and record the color you observe. For each solution, it is the first element in the formula that is responsible for color.

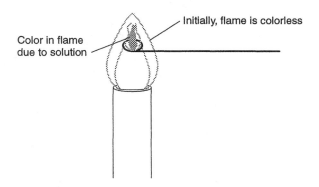

Color in flame due to solution

Initially, flame is colorless

Figure 5.2   Using a flame-test wire to test for flame color

Note: The color of potassium in the KCl flame is short-lived. Be sure to observe the color of the flame from the KCl solution within the first few seconds of heating. Repeat each flame test until you can describe the color of the flame produced. Clean the wire and repeat the flame test with the other solutions.

Identifying Solutions

Obtain unknown solutions as indicated by your instructor and record their code letters. Place 6–8 drops of each unknown solution in a clean spot plate. Use the flame-test procedure to determine the identity of the unknown solution. You may wish to recheck the flame color of the known solution that best matches the flame color of an unknown. For example, if you think your unknown is KCl, recheck the color of the KCl solution to confirm.

B. Electron Configuration

Write the electron configuration of each atom listed on the laboratory report. Indicate the number of valence electrons and the group number for the element.

C. Graphing a Periodic Property: Atomic Radius

The atomic radii for elements with atomic numbers 1–25 are listed in Table 5.1. On the graph, plot the atomic radius of each element against the atomic number of the element on the graph. Be sure to connect the points. Use the completed graph to answer questions in the report sheet about valence electrons and group number.

Table 5.1 Atomic Radii for the Elements with Atomic Numbers 1–25

Element	Symbol	Atomic Number	Atomic Radius (pm*)
first period			
hydrogen	H	1	37
helium	He	2	50
second period			
lithium	Li	3	152
beryllium	Be	4	111
boron	B	5	88
carbon	C	6	77
nitrogen	N	7	70
oxygen	O	8	66
fluorine	F	9	64
neon	Ne	10	70
third period			
sodium	Na	11	186
magnesium	Mg	12	160
aluminum	Al	13	143
silicon	Si	14	117
phosphorus	P	15	110
sulfur	S	16	104
chlorine	Cl	17	99
argon	Ar	18	94
fourth period			
potassium	K	19	231
calcium	Ca	20	197
scandium	Sc	21	160
titanium	Ti	22	150
vanadium	V	23	135
chromium	Cr	24	125
manganese	Mn	25	125

*(picometer = 10^{-12} m)

Report Sheet - Lab 5

Date _____   Name _____

Section _____   Team _____

Instructor _____   _____

Pre-Lab Study Questions
1. What is the color of a neon light?

2. Why does a sodium street lamp give off a different color light than a neon light?

3. What are the energy levels of electrons?

4. Why do some elements produce colorful flames?

Report Sheet - Lab 5

A. Flame Tests

Spot plate diagram

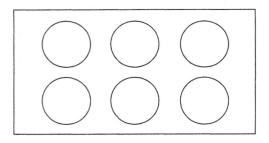

Solution	Element	Color of Flame
$CaCl_2$	Ca	_____
KCl	K	_____
$BaCl_2$	Ba	_____
$SrCl_2$	Sr	_____
$CuCl_2$	Cu	_____
NaCl	Na	_____

Unknown Solution(s)
Identification letter

☐   ☐   ☐

Color of flame   _____   _____   _____

Element present   _____   _____   _____

Questions and Problems

Q.1 You are cooking spaghetti in water you have salted with NaCl. You notice that when the water boils over, it causes the flame of the gas burner to turn bright orange. How would you explain the appearance of a color in the flame?

Report Sheet - Lab 5

B. Electron Configuration

Atom	Electron Configuration	Number of Valence Electrons	Group
O			
Na			
Ca			
Fe			8B
Zn			2B
Br			
Sr			
Cd			8B
Xe			
Cs			
Pb			
Ra			

Report Sheet - Lab 5

Questions and Problems

Q.2 Complete the following electron shells, subshells, and configurations:

Number of subshells in n = 3		Group number of carbon	
Number of orbitals in the 2p subshell		Subshell being filled by element with atomic number 47	
Number of electrons in 3d subshell		Subshell block that begins to fill after $4s^2$	
Number of electrons in a 3p orbital		Number of valence electrons in As	

Q.3 Give the symbol of the element that meets the following information:

Element that fills 3s sublevel		Element with five 3p electrons	
Period 4 element in the same group as F		$1s^2 2s^2 2p^6 3s^2 3p^6 4s^2 3d^{10} 4p^3$	
Element with $3d^6$		Element that completes n = 3	
Element with a half-filled 5p level		Period 6 element in the same group as Mg	

Report Sheet - Lab 5

C.   Graphing a Periodic Property: Atomic Radius

Atomic Radius vs. Atomic Number

**Atomic
radius (pm)**

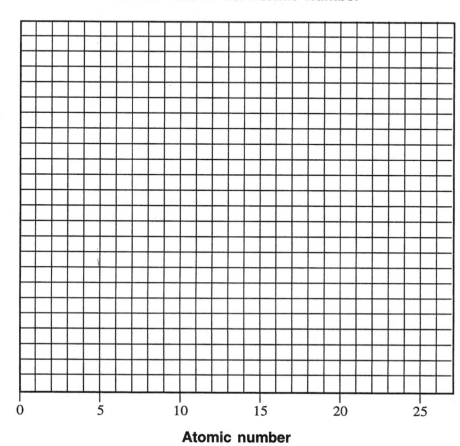

Atomic number

Questions and Problems

Q.4   Describe the change in the atomic radii for the elements in Period 2 from lithium to neon.

Q.5   Why does the change for the atomic radii of the elements in Period 3 from sodium to argon look similar to Period 2?

Nuclear Radiation

Goals

- Observe the use of a Geiger-Müeller radiation detection tube.
- Determine the effect of shielding materials, distance, and time of radiation.
- Complete a nuclear equation.

Discussion

Radioactivity occurs when a proton or neutron breaks down in the nucleus of an unstable atom or the particles in the nucleus are rearranged. Then a particle or energy called *nuclear radiation* is emitted from the nucleus. The nucleus has undergone *nuclear decay*. The most typical kinds of radiation include alpha particles (α), beta particles (β), and gamma rays (γ).

alpha decay $^{147}_{62}Sm \longrightarrow {}^{143}_{60}Nd + {}^{4}_{2}He$

alpha particle (α)

beta decay $^{40}_{20}Ca \longrightarrow {}^{40}_{21}Sc + {}^{0}_{-1}e$

beta particle (β)

gamma decay $^{167}_{68}Er \longrightarrow {}^{167}_{68}Er + {}^{0}_{0}\gamma$

gamma ray (γ)

If radiation passes through the cells of the body, the cells may be damaged. You can protect yourself by using shielding materials, by limiting the amount of time near radiation sources, and by keeping a reasonable distance from the radioactive source.

To detect radiation, a device such as a Geiger-Müeller tube is used. Radiation passes through the gas held within the tube, producing ion pairs. These charged particles emit bursts of current that are converted to flashes of light and audible clicks. In this experiment your teacher will demonstrate the use of the Geiger-Müeller tube to test the effects of shielding, time, and distance.

Lab Information

Time: 2 hr

Comments: Tear out the report sheets and place them beside the procedures.
               Follow your instructor's directions for protection from radiation.

Related Topics: Radioactivity, alpha particles, beta particles, gamma rays, shielding,
                nuclear decay, nuclear equations

Experimental Procedures

(This experiment will be done as a demonstration.)

 SAFETY GOGGLES PLEASE!

A. Background Count

Materials: Geiger-Müeller radiation detection tube

A.1 The level of radiation that occurs naturally is called *background radiation.* Set the radiation counter at the proper voltage for operating level. Let it warm up for 5 minutes. Remove all sources of radiation near the counter. Count any radiation present in the room by operating the counter for 1 minute. Record the counts. Repeat the background count for two more 1-minute intervals.

A.2 Total the counts in A.1, and divide by 3. This value represents the background level of radiation in counts per minute (cpm). This background count is the natural level of radiation that constantly surrounds and strikes us. For the rest of this experiment, the background radiation must be subtracted to give the radiation from the radioactive source alone.

B. Radiation from Radioactive Sources

Materials: Geiger-Müeller radiation detection tube, meterstick, 3–4 radioactive sources of alpha- and beta-radiation samples to test for radioactivity: Fiestaware™, minerals, old lantern mantles containing thorium compounds, camera lenses, old watches with radium-painted numbers on dials, smoke detectors containing Am-241, antistatic devices for records and film containing Po-210, some foods such as salt substitute (KCl), cream of tartar, instant tea, instant coffee, dry seaweed

Place one of the radioactive sources at a distance of 10–20 cm from the detection tube. Record the radiation emitted for 1 minute. Subtract the background count to obtain the radiation emitted by the source alone. Test other sources for 1 minute at the same distance.

C. Effect of Shielding, Time, and Distance

Materials: Geiger-Müeller radiation detection tube, meterstick, one of the radioactive sources used in part B, shielding materials: lead, paper, glass, cardboard, etc.

C.1 **Shielding** Place the radiation source at the same distance from the detection tube as it was in part B. Place various shielding materials (cardboard, paper, several pieces of cardboard, lead sheet, etc.) between the radioactive source and the detection tube. See Figure 6.1. Record the type of shielding used. Do not vary the distance of the source from the tube. For each type of shielding, record the radiation in counts per minute obtained from the source for 1 minute. Subtract the background count from each result to find the amount of radiation in cpm allowed by each type of shielding.

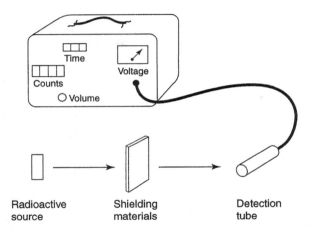

Figure 6.1 Measuring radiation using shielding

C.2 **Time** The more time you spend near a radioactive source, the greater the amount of radiation you receive. Keeping the distance constant, record the total counts for a radioactive source over time periods of 1, 2, and 5 minutes. Subtract background radiation from each to obtain the radiation from the source alone. For 2 and 5 minutes, subtract a background that is two times (2×) and five times (5×) the background count.

C.3 Use the results from C.2 to calculate the radiation received for 20 and 60 minutes.

C.4 **Distance** By doubling your distance from a radioactive source, you receive one-fourth (1/4) the intensity of the radiation. See Figure 6.2. Place a radioactive source 1 m (100 cm) from the tube. Record the counts for 1 minute at 100 cm. Decrease the distance of the radioactive source from the detection tube to 75 cm, 50 cm, 25 cm, and 10 cm. Stop measuring the radiation if the count becomes too great for the operating level of the counter. Record the counts for 1 minute at each distance. Subtract the background radiation from each.

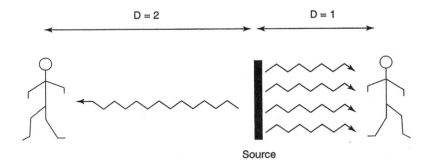

Figure 6.2 The effect of radiation lessens as the distance from the source increases

C.5 Calculate the ratio of the counts per minute at 50 and 100 cm. This will give the increase in radiation when the distance to the source is halved.

C.6 Graph the radiation level (cpm) against the distance of the source from the detection tube. Review instructions for preparing a graph found in the preface.

C.7 Using the graph, predict the counts per minute that would be obtained at distances of 20 and 40 cm.

Report Sheet - Lab 6

Date _____   Name _____

Section _____   Team _____

Instructor _____   _____

Pre-Lab Study Questions

1. In what part of the atom do alpha or beta particles originate?

2. Why is protection from radiation needed?

3. What are some medical uses of radiation?

A. Background Count

A.1 Counts during   minute 1   _____

                       minute 2   _____

                       minute 3   _____

A.2 Total counts   _____

                   Average background count   _____ counts/minute (cpm)

B. Radiation from Radioactive Sources

Item Tested	Counts per Minute (cpm)	Background (cpm)	Source (cpm)	Type of Radiation

Questions and Problems

Q.1 Which item was the most radioactive?

Q.2 If that item is a consumer product, what is the source and purpose of the radioactive material?

C. Effect of Shielding, Time, and Distance

C.1 **Shielding**

Shielding	Counts per Minute	Background (cpm)	Source (cpm)
No shielding			

Questions and Problems

Q.3 Which type of shielding provides the best protection from radiation? Why?

Report Sheet - Lab 6

C.2 Time

Minutes	Counts	Background (cpm × no. of minutes)	Source
1			
2			
5			
C.3 20 (calculated)			
60 (calculated)			

C.4 Distance

Distance	Counts per Minute	Background (cpm)	Source (cpm)
100 cm			
75 cm			
50 cm			
25 cm			
10 cm			

C.5 Ratio: $\dfrac{\text{Counts/min at 50 cm}}{\text{Counts/min at 100 cm}} = \underline{\hspace{2cm}}$

Questions and Problems

Q.4 Complete the nuclear equations by filling in the correct symbols:

$$^{27}_{13}\text{Al} + \boxed{} \longrightarrow\ ^{24}_{11}\text{Na} +\ ^{4}_{2}\text{He}$$

$$^{131}_{53}\text{I} \longrightarrow\ ^{0}_{-1}e + \boxed{}$$

$$^{96}_{40}\text{Zr} + \boxed{} \longrightarrow\ ^{1}_{0}\text{n} +\ ^{99}_{42}\text{Mo}$$

Report Sheet - Lab 6

C.6 **Graph**

Counts per minute vs. distance from source

Counts/minute

0

Distance (cm) from source

C.7 Use your graph to estimate radiation levels at the following distances from the source:

Estimated cpm at 40 cm = ———————

Estimated cpm at 20 cm = ———————

Questions and Problems

Q.5 Write the symbols for the following types of radiation:

a. alpha particle _____   b. beta particle _____   c. gamma ray _____

Q.6 List some shielding materials adequate for protection from:

a. alpha particles _____

b. beta particles _____

c. gamma rays _____

Q.7 The bacteria in some foods are sterilized by placing the food near a source of ionizing radiation. Does that mean that the food becomes radioactive? Explain.

7

Compounds and Their Formulas

Goals

- Compare physical properties of a compound with the properties of the elements that formed it.
- Identify a compound as ionic or covalent.
- Determine the subscripts in the formula of a compound.
- Write the electron-dot structure for an atom and an ion.
- Write a correct formula and name of an ionic or covalent compound.
- Write a correct formula and name of a compound containing a polyatomic ion.

Discussion

Nearly everything is made of compounds. A compound consists of two or more different elements that are chemically combined. Most atoms form compounds by forming octets in their outer shells. The attractions between the atoms are called *chemical bonds*. For example, when a metal combines with a nonmetal, the metal loses electrons to form a positive ion and the nonmetal gains electrons to form a negative ion. The attraction between the positive ions and the negative ions is called an *ionic bond*. When two nonmetals form a compound, they share electrons and form *covalent bonds*. In covalent compounds, the atoms are bonded as individual units called *molecules*. See Table 7.1.

Table 7.1   *Types of Bonding in Compounds*

Compound	Types of Elements	Characteristics	Type of Bonding
NaCl	Metal, nonmetal	Ions (Na^+, Cl^-)	Ionic
$MgBr_2$	Metal, nonmetal	Ions (Mg^{2+}, Br^-)	Ionic
CCl_4	Two nonmetals	Molecules	Covalent
NH_3	Two nonmetals	Molecules	Covalent

In a compound, there is a definite proportion of each element. This is represented in the formula, which gives the lowest whole number ratio of each kind of atom. For example, water has the formula H_2O. This means that two atoms of hydrogen and one atom of oxygen are combined in every molecule of water. Water never has any other formula.

When we observe a compound or an element, we see physical properties such as color and luster. We measure other physical properties such as density, melting point, and boiling point. When elements undergo chemical combination, the physical properties change to the physical properties of the new substances that form. For example, when silver tarnishes, the physical property of the shiny, silver metal changes to the dull, gray color as silver combines with sulfur to form tarnish, Ag_2S. A chemical change has occurred when the reaction between elements causes a change in their physical properties.

A.   Electron-Dot Structures

When atoms of metals in Groups 1, 2, or 3 react with atoms of nonmetals in Groups 5, 6, or 7, the metals lose electrons and the nonmetals gain electrons in their valence shells. We can predict the number of electrons lost or gained by analyzing the electron-dot structures of the atoms. In an electron-dot structure, the valence electrons are represented as dots around the symbol of the atom. For example, calcium, electron arrangement 2-8-8-2, has two valence electrons and an electron-dot structure with two dots. Chlorine, electron arrangement 2-8-7, has seven valence electrons and an electron-dot structure with seven dots.

$$\overset{\displaystyle\cdot}{Ca}\cdot \qquad\qquad \overset{\displaystyle\cdot\cdot}{\underset{\displaystyle\cdot\cdot}{\vphantom{|}\textbf{:}\,Cl\,\cdot}}$$

Ca loses two electrons to attain an octet. This gives it an ionic charge of 2+. It is now a calcium ion with an electron arrangement of 2-8-8. As a positive ion, it keeps the same name as the element.

	Calcium Atom, Ca	**Calcium Ion, Ca^{2+}**	
Electron arrangement	2-8-8-2	2-8-8	(Two electrons lost)
Number of protons	$20p^+$	$20p^+$	(Same)
Number of electrons	$20e^-$	$18e^-$	(Two fewer electrons)
Net ionic charge	0	2+	

When nonmetals (5, 6, or 7 valence electrons) combine with metals, they gain electrons to become stable, and form negatively charged ions. For example, a chlorine atom gains one valence electron to become stable with an electron arrangement of 2-8-8. With the addition of one electron, chlorine becomes a chloride ion with an ionic charge of 1–. In the name of a binary compound with two different elements, the name of the negative ion ends in *ide*.

	Chlorine Atom, Cl	**Chloride Ion, Cl$^-$**	
Electron arrangement	2-8-7	2-8-8	(Electron added)
Number of protons	$17p^+$	$17p^+$	(Same)
Number of electrons	$17e^-$	$18e^-$	(One more electron)
Net ionic charge	0	1–	

B. Ionic Compounds and Formulas

The group number on the periodic table can be used to determine the ionic charges of elements in each family of elements. *Nonmetals form ions when they combine with a metal.*

Group number	*1*	*2*	*3*	*4*	*5*	*6*	*7*	*8*
Valence electrons	$1e^-$	$2e^-$	$3e^-$	$4e^-$	$5e^-$	$6e^-$	$7e^-$	$8e^-$
Electron change	lose 1	lose 2	lose 3	none	gain 3	gain 2	gain 1	no change
Ionic charge	1+	2+	3+	none	3–	2–	1–	none

In an ionic formula, the *number of electrons lost is equal to the number of electrons gained.* The overall net charge is zero. To balance the charge, we must determine the smallest number of positive and negative ions that give an overall charge of zero (0). We can illustrate the process by representing the ions Ca^{2+} and Cl$^-$ as geometric shapes.

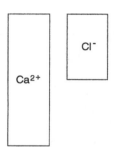

The charge is balanced by using two Cl^- ions to match the charge of the Ca^{2+} ion. The number of ions needed gives the subscripts in the formula for the compound $CaCl_2$. (The subscript 1 for Ca is understood.) In any ionic formula, *only the symbols are written, not their ionic charges.*

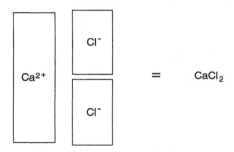

C. Ionic Compounds with Transition Metals

Most of the transition metals can form more than one kind of positive ion. We will illustrate variable valence with iron. Iron forms two ions, one (Fe^{2+}) with a 2+ charge, and another (Fe^{3+}) with a 3+ charge. To distinguish between the two ions, a Roman numeral that gives the ionic charge of that particular ion follows the element name. The Roman numeral is always included in the names of compounds with variable positive ions. In an older naming system, the ending *ous* indicates the lower valence; the ending *ic* indicates the higher one. See Table 7.2.

Table 7.2 *Some Ions of the Transition Elements*

Ion	Names	Compound	Names
Fe^{2+}	Iron(II) ion or ferrous ion	$FeCl_2$	Iron(II) chloride or ferrous chloride
Fe^{3+}	Iron(III) ion or ferric ion	$FeCl_3$	Iron(III) chloride or ferric chloride
Cu^+	Copper(I) ion or cuprous ion	$CuCl$	Copper(I) chloride or cuprous chloride
Cu^{2+}	Copper(II) ion or cupric ion	$CuCl_2$	Copper(II) chloride or cupric chloride

Among the transition metals, a few elements (zinc, silver, and cadmium) form only a single type of ion; they have a fixed ionic charge. Thus, they are *not* variable and *do not need* a Roman numeral in their names.

Zn^{2+} zinc ion Ag^+ silver ion Cd^{2+} cadmium ion

D. Ionic Compounds with Polyatomic Ions

A compound that consists of three or more kinds of atoms will contain a *polyatomic ion*. A polyatomic ion is a group of atoms with an overall charge. That charge, which is usually negative, is the result of adding electrons to a group of atoms to complete octets. The most common polyatomic ions consist of the nonmetals C, N, S, P, Cl, or Br combined with two to four oxygen atoms. Some examples are given in Table 7.3. The ions are named by replacing the ending of the nonmetal with *ate* or *ite*. The *ite* ending has one oxygen less than the most common form of the ion, which has an *ate* ending. Ammonium ion, NH_4^+, is positive because its group of atoms lost one electron.

Table 7.3 *Some Polyatomic Ions*

Common Polyatomic Ion		**One Oxygen Less**	
NH_4^+	ammonium ion		
OH^-	hydroxide ion		
NO_3^-	nitrate ion	NO_2^-	nitrite ion
CO_3^{2-}	carbonate ion		
HCO_3^-	bicarbonate ion (hydrogen carbonate ion)		
SO_4^{2-}	sulfate ion	SO_3^{2-}	sulfite ion
HSO_4^-	bisulfate ion (hydrogen sulfate ion)	HSO_3^-	bisulfite ion (hydrogen sulfite ion)
PO_4^{3-}	phosphate ion	PO_3^{3-}	phosphite ion

To write a formula with a polyatomic ion, we determine the ions needed for charge balance just as we did with the simple ions. When two or more polyatomic ions are needed, the formula of the ion is enclosed in parentheses and the subscript placed *outside. No change is ever made in the formula of the polyatomic ion itself.* Consider the formula of the compound formed by Ca^{2+} and NO_3^- ions.

E. Covalent (Molecular) Compounds

Covalent bonds form between two nonmetals found in Groups 4, 5, 6, or 7 or H. In a *covalent compound,* octets are achieved by sharing electrons between atoms. The sharing of one pair of electrons is called a single bond. A double bond is the sharing of two pairs of electrons between atoms. In a triple bond, three pairs of electrons are shared. To write the formula of a covalent compound, determine the number of electrons needed to complete an octet. For example, nitrogen in Group 5 has five valence electrons. Nitrogen atoms need three more electrons for an octet; they share three electrons.

Electron-Dot Structures

The formulas of covalent compounds are determined by sharing the valence electrons until each atom has an octet. For example, in water (H_2O), oxygen shares two electrons with two hydrogen atoms. Oxygen has an octet and hydrogen is stable because it has two electrons in the first valence shell.

Dot Structure for H_2O

In another example, we look at a compound, CO_2, that has double bonds. In the elements' electron-dot structures, carbon has 4 valence electrons and each oxygen atom has 6. Thus a total of 16 (4 + 6 + 6) electrons can be used in forming the octets by sharing electrons. We can use the following steps to determine the electron-dot structure for CO_2:

1. Connect the atoms with pairs of electrons, thus using 4 electrons.

2. Place the remaining 12 electrons (16 − 4) around the atoms. Don't add more electrons.

3. If octets *cannot* be completed, try sharing more electrons. In step 2, the octets are complete for the oxygen atoms, but not for the carbon. One pair of electrons from each oxygen atom is moved to share with carbon. Now all the atoms have octets. There are still 16 electrons used, but they are now arranged to give each atom an octet. There are two double bonds in the CO_2 molecule.

sharing two pairs of electrons makes double bonds

Names of Covalent Compounds

Binary (two-element) covalent compounds are named by using *prefixes* that give the number of atoms of each element in the compound. The first nonmetal is named by the element name; the second ends in *ide*. The prefixes are derived from the Greek names: mono (1), di (2), tri (3), tetra (4), penta (5), hexa (6), hepta (7), and octa (8). Usually the prefix *mono* is not shown for the first element. See Table 7.4.

Table 7.4 *Some Formulas and Names of Covalent Compounds*

Formula	Name
CO	carbon **mono**xide
CO_2	carbon **di**oxide
PCl_3	phosphorus **tri**chloride
N_2O_4	**di**nitrogen **tetr**oxide (drop *a* in a double vowel)
SCl_6	sulfur **hexa**chloride

F. Electron Dot Structures and Molecular Shape

A molecule has a shape such as linear, bent, trigonal planar, pyramidal, or tetrahedral. The electron-pair repulsion (VSEPR) model indicates that the bond angles in a molecule are determined when the valence electrons in bonds and lone pairs move as far apart as possible. Counting the groups of electrons that are shared pairs and lone pairs determines the electron geometry and bond angle. Atoms attached to the groups of electrons determine the molecular geometry. See Table 7.5.

Table 7.5 *Molecular Geometry for Atoms*

Total electron groups	Shared pairs	Lone pairs	Electron geometry	Bond angle	Molecular geometry	Examples
2	2	0	linear	180°	linear	$BeCl_2$, CO_2
3	3	0	trigonal planar	120°	trigonal planar	BF_3
3	2	1	trigonal planar	120°	bent	SO_2
4	4	0	tetrahedral	109°	tetrahedral	CH_4, CCl_4
4	3	1	tetrahedral	109°	pyramidal	NH_3
4	2	2	tetrahedral	109°	bent	H_2O

Polar and Nonpolar Molecules

When there is a difference of 0.4 or more between the electronegativity of the central atom and an attached atom, the bond is polar. If the polar bonds in a molecule are symmetrical and the dipoles cancel, the molecule itself is nonpolar. When the dipoles do not cancel, the molecule is polar.

Lab Information

Time: 2–3 hr

Comments: Tear out the report sheets and place them beside the matching procedures.

Related Topics: Ions, ionic bonds, naming ionic compounds, covalent bond, covalent compounds, naming covalent compounds

Experimental Procedures

 **GOGGLES REQUIRED!**

A.   Electron-Dot Structures

Write the electron arrangements for atoms and their ions. Determine the number of electrons lost or gained and write the electron-dot structure of the ion that each would form along with its symbol, ionic charge, and name.

B.   Ionic Compounds and Formulas

Materials: Reference books: *Merck Index* or *CRC Handbook of Chemistry and Physics,* display of compounds

B.1 **Physical properties** In the laboratory display of compounds observe NaCl, sodium chloride. Describe its appearance. Using a chemistry reference such as the *Merck Index* or the *CRC Handbook of Chemistry and Physics,* record the density and melting point.

B.2 **Formulas of ionic compounds** Use the periodic table to write the positive and negative ion in each compound. Use charge balance (net total = zero) to write the correct formula. Use subscripts when two or more ions are needed.

B.3 **Names of ionic compounds** From the formula of each ionic compound, write the compound name by placing the metal name first, then the nonmetal name ending in *ide.*

C.   Ionic Compounds with Transition Metals

C.1 **Physical properties** In the display of compounds observe $FeCl_3$, iron(III) chloride or ferric chloride. Describe its appearance. Using a chemistry reference such as the *Merck Index* or the *CRC Handbook of Chemistry and Physics,* record the density and melting point.

C.2 **Formulas of ionic compounds** Use the periodic table to write the positive and negative ion in each compound. Use charge balance (net total = zero) to write the correct formula. Use subscripts when two or more ions are needed.

C.3 **Names of ionic compounds** From the formula of each ionic compound, write the compound name by placing the metal name first, then the nonmetal name ending in *ide.* Be sure to indicate the ionic charge if the transition metal has a variable valence by using a Roman numeral or using the *ous* or *ic* ending.

D.   Ionic Compounds with Polyatomic Ions

D.1 **Physical properties** In the display of compounds observe K_2CO_3, potassium carbonate. Describe its appearance. Using a chemistry reference book such as the *Merck Index* or the *CRC Handbook of Chemistry and Physics,* record the density and melting point.

D.2 **Formulas of ionic compounds** Use the periodic table to write the positive and negative (polyatomic) ion in each compound. Use charge balance (net total = zero) to write the correct formula. Use subscripts when two or more ions are needed. Use parentheses when two or more polyatomic ions are needed for charge balance.

D.3 **Names of ionic compounds** Name the compounds listed, using the correct names of the polyatomic ions.

E. Covalent (Molecular) Compounds

E.1 **Electron-dot formulas of elements** Write the electron-dot structure for each nonmetal.

E.2 **Physical properties** In the display of compounds observe water, H_2O. Describe its appearance. Using a chemistry reference such as the *Merck Index* or the *CRC Handbook of Chemistry and Physics,* record the density and melting point.

E.3 **Electron-dot structures** Write the electron-dot structure for each covalent compound. Name each compound, using prefixes to indicate the number of atoms of each element. By convention, the prefix *mono* can be omitted from the name of the first nonmetal.

F. Electron Dot Structures and Molecular Shape

Obtain a molecular model set and build a model of each of the molecules or ions listed in the report sheet.

Complete the following for each of the molecules or polyatomic ions listed in the report page.

1. Draw the electron dot structure.

2. Count the electron groups around the central atom.

3. Use VSEPR to determine the electron geometry.

4. Determine the bond angle.

5. Count the numbers of atoms bonded to the central atom.

6. Use the number of bonded atoms to identify the molecular geometry.

7. Indicate if the molecules listed would be polar or nonpolar.

Energy and Specific Heat

Goals

- Distinguish between a calorie, kilocalorie, and nutritional Calorie.
- Use the specific heat of water to calculate heat lost or gained.
- Calculate the specific heat in cal/g °C and J/g °C of a metal object.
- Calculate the caloric values of foods in kcal/g to calculate the kilocalories in a serving of food.
- Use nutrition data on food products to determine the kilocalories in one serving.

Discussion

A.   Specific Heat of a Metal

Every substance has the capacity to absorb heat. When heat is added to a substance, the temperature of that substance increases. Different substances vary in the amount of heat required to raise their temperatures by 1°C. **Specific heat** is the amount of heat in calories or joules that raises the temperature of 1 g of a substance by 1°C.

$$\text{Specific heat } = \frac{\text{Amount of heat (cal or J)}}{\text{Mass(g)} \times \text{change in temperature } \Delta T \ (°C)}$$

Water has a very large specific heat, 1.00 cal/g °C or 4.18 J/g °C, compared to most substances. Water requires more heat than other substances before its temperature increases by 1°C. Because the specific heat of a substance is unique for that substance, specific heat can be used to identify a particular substance.

Table 8.1 Specific Heat Values for Selected Substances		
Substance	Specific Heat	
	cal/g °C	J/g °C
Water, H_2O (liquid)	1.00	4.18
Iron	0.11	0.46
Copper	0.093	0.39
Aluminum	0.22	0.92
Lead	0.031	0.13

In this experiment, an insulated Styrofoam cup is used as a calorimeter (Figure 8.1). A measured amount of water is placed in the cup. After a metal is heated to the temperature of boiling water (about 100°C), it is quickly transferred to the water in the calorimeter. The heat lost by the hot metal is used to warm the water. As the temperature of the water increases, the temperature of the metal decreases until both the water and the metal reach the same final temperature. By measuring the increase in the temperature of the water, we can calculate the amount of heat (calories or joules) given off by the hot metal.

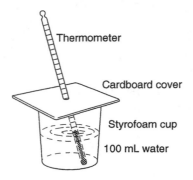

Figure 8.1 A calorimeter consists of a Styrofoam cup(s), cover, water, and a thermometer

For example, suppose 100. g of water in the Styrofoam cup has an initial temperature of 24.5°C, and a final temperature of 27.0°C. The specific of water is 1.00 cal/g °C and the temperature change is 2.5°C (27.0°C – 24.5°C).

Step 1: Calculate the heat gain for the water

Heat (cal)   = g x ΔT (°C) x specific heat

Heat (cal)   = 100. g x 2.5°C x 1.00 cal/g °C

= 250 cal

Because the heat gained by the water sample is equal to the heat lost by the metal, we know that the metal lost 250 calories of heat. If the boiling water that heated the metal had a temperature of 100.0°C, the temperature change for the metal is 73.0°C (100.0 – 27.0°C). Using the heat loss of the metal, and a mass (g) of the metal of 32.3 g, and the decrease in the temperature of the metal, we can calculate the specific heat of that metal.

Step 2: Calculate the specific heat of the metal.

$$\text{Specific heat (metal)} \quad = \quad \frac{\text{cal}}{\text{Mass (g) x }\Delta\text{T (°C)}} \quad = \quad \frac{250 \text{ cal}}{(32.3 \text{ g})(73.0°C)}$$

= 0.11 cal/g°C

The metal is probably iron because the experimental result matches the known specific heat of iron.

B. Measuring the Caloric Value of a Food

The calories in a sample of food are measured in the laboratory using a calorimeter. In this lab, you will use a simple setup to illustrate how the calories in a sample of food can be determined. Burning a cheese puff or some chips releases heat that is used to heat water in an aluminum can. In our setup, some of the heat is lost to the can and the surrounding air. However, you can get an idea of how nutritionists determine the caloric content of foods. Food calories are usually given as kilocalories (Calories). Caloric values for foods are reported as kcal/g of food.

C. Food Calories

Our diets contain foods that provide us with energy. We need energy to make our muscles work, to breathe, to synthesize molecules in the body such as protein and fats, and to repair tissues. A typical diet required by a 25-year-old woman is about 2000–2500 kcal. By contrast a bicycle rider has a higher energy requirement and needs a diet that provides 4000 kcal. The nutritional energy of food is determined in Calories (Cal), which are the same as 1000 cal or 1 kilocalorie.

Nutritionists use calorimeters to establish the caloric values for the three food types: carbohydrates, 4 kcal/g; fats, 9 kcal/g; and proteins, 4 kcal/g. By measuring the amount of each food type in a serving, the kilocalories can be calculated. For example, a candy that is composed of 12 g of carbohydrate will provide 48 kcal. Usually the values are rounded to the nearest tens place.

$$12 \text{ g carbohydrate} \times \frac{4 \text{ kcal}}{1 \text{ g carbohydrate}} = 48 \text{ kcal or } 50 \text{ kcal}$$

Lab Information

Time:	2–3 hr
Comments:	Tear out the lab report sheets and place them beside the matching procedures. Be careful with boiling water.
Related topics:	Specific heat, measuring heat energy, calculating heat in calories and joules, nutritional calorie, caloric values.

Experimental Procedures    **GOGGLES REQUIRED!**

A. Specific Heat of a Metal

> **Materials:** Thermometer, Bunsen burner, ring stand, iron ring, wire screen, 400-mL beaker, balance, calorimeter (Styrofoam cup and cover), stirring rod, and metal object

Fill the 400-mL beaker about two-thirds full of water. Place the beaker and water on a wire screen set on an iron ring so that the ring is about 6 cm above the top of the Bunsen burner. Light your Bunsen burner and begin heating the water in the beaker.

A.1 Obtain a metal object. Record any identification number for the metal. Using a balance, determine the mass of the metal object. Record. Tie a length of string or fishing line to the metal object and gently lower it into the water bath. Allow the water bath containing the metal object to boil for 10 minutes.

A.2 Obtain a Styrofoam cup and determine the mass of your calorimeter. Record.

A.3 Add about 50 mL of water to the Styrofoam cup (calorimeter) and determine the combined mass. Record. Note: There must be enough water to cover the metal object. If the metal object is large, you may need to place up to 100 mL of water in the Styrofoam cup.

A.4 After the water in the water bath has boiled for 10 minutes or more, use a thermometer to measure the temperature of the boiling water. Record. This is also the initial temperature of your metal object.

A.5 Measure the temperature of the water in your Styrofoam cup (calorimeter). Record.

A.6 Carefully remove the metal object from the text tube (boiling water bath) and quickly place the hot metal object in the water in the Styrofoam cup (calorimeter). Place the cover on top of the calorimeter, and use the thermometer or a stirring rod to gently stir. Be careful not to hit the metal object with the thermometer. The highest temperature reached will be the final temperature of both the water and the metal object. Record this temperature.

Dry off the metal object. If you are going to run a second trial, repeat the preceding procedures.

Calculations

A.7 Calculate the temperature change for the water in the Styrofoam calorimeter.
Temperature change for the water = Final temperature of water – initial temperature of water

A.8 Calculate the heat in calories that were transferred from the metal to heat the water.
$$cal = g \times \Delta T \, (°C) \times \text{specific heat}$$

A.9 State the number of calories lost by the metal object.
Heat (cal) lost by the metal = Heat (cal) gained by the water

A.10 Calculate the temperature change for the metal.
$\Delta T \, (°C)$ of metal =
Initial temperature of the metal (boiling water bath) – final temperature (metal and water)

A.11 Calculate the specific heat of the metal
$$\text{Specific heat of the metal} = \frac{\text{amount of heat (calories)}}{\text{Mass of metal (g) x temperature change } (\Delta T \, °C)}$$

Calculate the specific heat of the metal in joules/g °C.

A.12 Use Table 8.1 to try to identify the type of metal in the object. Your instructor may add other values for identification.

B. Measuring the Caloric Value of a Food
This may be an instructor demonstration

> **Materials:** Aluminum can, food sample such as chips or cheese puffs, thermometer, two iron rings, a clamp, wire screen, Bunsen burner

B.1 Obtain an aluminum can and determine its mass. Add about 100 mL of water and determine the combined mass of the water and the can.

> ***Set up in the hood:*** Place the can in an iron ring that holds it. Attach a second iron ring covered with a wire screen a short distance below the aluminum can. Suspend a thermometer from a clamp so that the bulb is below the water level in the aluminum can.

B.2 Weigh the food sample. Record.

B.3 Record the initial temperature of the water in the can. Place the food sample on the wire screen below the aluminum can. Ignite the food sample using a match or Bunsen burner. Remove the heat source immediately and let the food sample burn. Record the final (highest) temperature reached by the water in the can.

B.4 Weigh the ash and any remaining food sample. Record

Calculations

B.5 Calculate the mass of water in the aluminum can.

B.6 Calculate the temperature change of the water after it is heated by the combustion reaction of the burning food.

B.7 Calculate the heat gain of the water in the aluminum can in calories and kilocalories. This is also the heat lost by the food sample.

Heat (cal) = mass (g) water $\times$ ΔT (°C) $\times$ 1.00 cal/g °C

Kilocalories (Calories) = heat(cal) $\times$ 1 kcal/1000 cal

B.8 Calculate the mass of food sample that burned. Subtract the mass of ash and any remaining unburned food after combustion.

B.9 Calculate the caloric value of the food.

$$\text{Caloric value} \quad = \quad \frac{\text{kcal (Cal)}}{\text{Mass (g) of food sample burned}}$$

C. Food Calories

Materials: Food products with nutrition data on labels

C.1 Obtain a food product that has a Nutrition Facts label. Indicate the serving size.

C.2 List the grams of fat, carbohydrate, and protein in one serving of the food.

C.3 From the mass of each food type, calculate the Calories (kcal) of each food type in one serving using the accepted caloric values.

C.4 Determine the total Calories (kcal) in one serving.

C.5 Compare your total to the Calories listed on the upper portion of the label. Usually these totals are rounded to the nearest tens place.

Energy and States of Matter

Goals

- Prepare a heating curve and a cooling curve.
- Use the specific heat of water to calculate heat lost or gained.
- Calculate the heat of fusion for water.

Discussion

A. A Heating Curve for Water

The temperature of a substance indicates the kinetic energy (energy of motion) of its molecules. When water molecules gain heat energy, they move faster and the temperature rises. Eventually the water molecules gain sufficient energy to separate from the other liquid molecules. The liquid changes to a gas in a change of state called *boiling*. The change of state from liquid to gas is indicated when the water temperature becomes constant. It is more obvious when a graph is drawn of the temperature change of the substance that is heated. When a liquid boils, a horizontal line (*plateau*) appears on the graph, as shown in Figure 9.1. This constant temperature is called its *boiling point*.

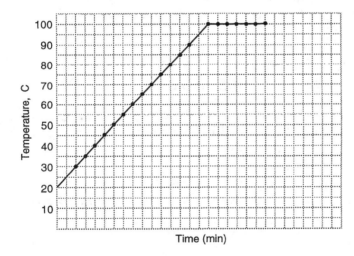

Figure 9.1 Example of a graph of a heating curve

B. Graphing a Cooling Curve

When a liquid cools, its particles move more slowly. Attractions between the particles become so strong that they form a solid. This change of state is called freezing. There is no change in the temperature while the liquid is freezing. When the liquid reaches it freezing point, its temperature becomes constant. On the graph in Figure 9.2, the drop in temperature is shown. As the liquid cools, the temperature may drop temporary below its freezing point. But as it continues to freeze, you will notice a jump up in the temperature. This condition is called supercooling. The horizontal line or plateau that appears after the supercooling indicates the freezing point of the substance as it changes from a liquid to a solid.

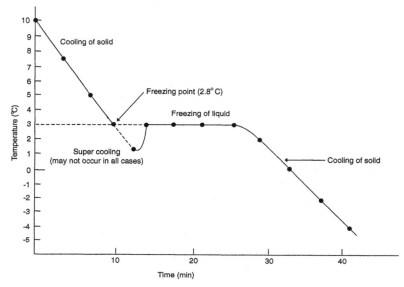

Figure 9.2 A typical cooling curve showing supercooling and the freezing point plateau.

C. Energy in Changes of State

Heat of Fusion

Changing state from solid to liquid (melting) requires energy. Ice melts at 0°C, a constant temperature. At that melting point, the amount of heat required to melt 1 g of ice is called the *heat of fusion*. For water, the energy needed to melt 1 g of ice (0°C) is 80. calories. That is also the amount of heat released when liquid water freezes to solid ice at 0°C.

Melting (0°C): $H_2O(s)$ + $heat_{fusion}$ (80. cal/g) $\longrightarrow$ $H_2O(l)$

Freezing (0°C): $H_2O(l)$ $\longrightarrow$ $H_2O(s)$ + $heat_{fusion}$ (80. cal/g)

In this experiment, ice will be added to a sample of water. From the temperature change, the amount of heat lost by the water sample can be calculated. This is also the amount of heat needed to melt the ice.

Heat (cal) lost by water = heat (cal) gained to melt ice
= g water × ΔT × 1.00 cal/g °C

By measuring the amount of ice that melted, the heat of fusion can be calculated as follows:

$$\text{Heat of fusion (cal/g)} = \frac{\text{heat (cal) gained to melt ice}}{\text{grams of ice}}$$

Heat of Vaporization

A similar situation occurs for a substance that changes from a liquid to a gas (vapor). Water boils at 100°C, its boiling point. At that temperature, the energy required to convert liquid to gas is called the *heat of vaporization*. For water, the energy required to vaporize 1 g of water at 100°C is 540 calories. This is also the amount of heat released when 1 gram of steam condenses to liquid at 100°C.

Boiling (100°C): $H_2O(l)$ + $heat_{vaporization}$ (540 cal/g) $\longrightarrow$ $H_2O(g)$

Condensation (100°C): $H_2O(g)$ $\longrightarrow$ $H_2O(l)$ + $heat_{vaporization}$ (540 cal/g)

Lab Information

Time: 2 hr
Comments: Tear out the report sheets and place them beside the matching procedures.
                Be careful with boiling water.
                Use mitts or beaker tongs to move hot beakers, or let them cool.
Related Topics: Changes of state, heating and cooling curves, heat of fusion, heat of vaporization

Experimental Procedures

 **GOGGLES REQUIRED!**

A.   A Heating Curve for Water

Materials: Beaker (250- or 400-mL), Bunsen burner (or hot plate), ring stand, graduated
cylinder, iron ring, wire gauze, clamp, thermometer, timer

Using a graduated cylinder, pour 100 mL of cool water into a 250-mL beaker. As shown in Figure
9.3, place the beaker on a hot plate or on a wire screen placed on an iron ring above a Bunsen
burner. The height of the iron ring should be about 3–5 cm above the burner. Tie a string to the
loop in the top of the thermometer or place the thermometer securely in a clamp. Adjust the thermometer so that the bulb is in the center portion of the liquid. (Do not let the thermometer rest on
the side or bottom of the beaker.)

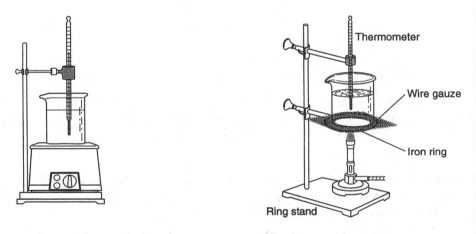

Figure 9.3   Setup for heating water with (left) a hot plate or (right) a Bunsen burner

A.1   Measure and record the initial temperature of the water. Light the burner (or use a hot plate).
      Using a timer or a watch with a second hand, record the temperature of the water at 1-minute
      intervals. Eventually the water will come to a *full boil*. (The early appearance of small bubbles of
      escaping gas does not indicate boiling.) When the water is boiling, the temperature has become
      constant. Record the boiling temperature for another 4–5 minutes.

A.2   Prepare a heating curve for water by graphing the temperature versus the time. Review graphing
      data found in the preface. Label the parts of the graph that represent the liquid state and boiling.

A.3   The plateau (flat part of graph) indicates the *boiling point* of the water. Record its value.

A.4   Calculate the temperature change needed for the cool water to reach the boiling point (plateau).

A.5   Using the measured volume of water and its density (1.00 g/mL), calculate the mass of the water.

A.6   Calculate the heat in calories used to heat the water to the boiling point.

      Calories = mass × ΔT × 1.00 cal/g °C

B. Graphing a Cooling Curve

Materials: The freezing-point apparatus, which consists of a large test tube containing a small amount of Salol (phenylsalicylate) and fitted with a two-hole stopper, thermometer, and wire stirrer. Do not try to pull out the stopper or thermometer. It is frozen in the Salol until heated. Hot-water bath: 400-mL beaker about $^1/_2$ full of water, Bunsen burner, iron ring, and wire screen

The substance called Salol is a solid at room temperature. Your instructor will indicate the location of the already prepared test tubes containing Salol and stirring apparatus. See Figure 9.4.

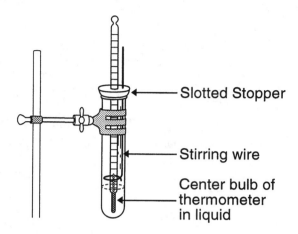

Slotted Stopper

Stirring wire

Center bulb of thermometer in liquid

Figure 9.4   Stirring apparatus for cooling curve and freezing point determination of Salol.

B.1   The Salol in the freezing-point apparatus must be melted before this experiment can begin. Prepare a hot water bath and place the test tube setup containing the Salol in the warm water. Let the temperature of the melted Salol go up to about 70°C *not to the boiling temperature of water*. All the Salol should now be liquid.

Turn off the Bunsen burner and remove the test tube with the Salol from the warm water. Clamp the test tube and contents to a ring stand. Gently raise and lower the wire stirrer to mix the contents. Each minute record the temperature as the liquid Salol cools. You will have to stop stirring when the stirrer becomes frozen in the Salol. After solid forms, take at least five more temperature readings. A constant temperature for five or more minutes indicates that the Salol has reached its freezing point. Return the freezing-point apparatus and contents to your instructor.

B.2   Plot the cooling curve for Salol on the graph provided in the report page. For a review of graphing, see page xviii in the preface of this lab manual. Label the areas of liquid state and solid state, super-cooling (if any), and the freezing point of the Salol.

C. Energy in Changes of State

Materials: Calorimeter (Styrofoam® cup and cardboard cover), thermometer,
50- or 100-mL graduated cylinder, 100-mL beaker, ice

C.1 Weigh an empty Styrofoam cup.

C.2 Add 100 mL of water to the cup and reweigh.

C.3 Record the initial temperature of the water in the calorimeter. See Figure 9.5. Add 2 or 3 ice cubes (or crushed ice that fills a 100-mL beaker) to the water in the cup. Stir strongly. Check the temperature of the ice water. Add ice until the temperature drops to 2–3°C. If some ice is not melted, remove it immediately. Record the final temperature of the water.

C.4 Weigh the Styrofoam (calorimetry) cup with the initial sample of water and the melted ice. The increase in mass indicates the amount of ice that melted.

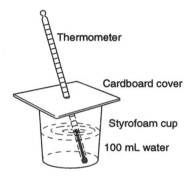

Thermometer

Cardboard cover

Styrofoam cup

100 mL water

Figure 9.5 Calorimetry setup with water, a thermometer, Styrofoam cup, and a cardboard cover

Calculations

C.5 Calculate the mass of water added to the Styrofoam cup.

C.6 Calculate the temperature change (ΔT) for the water.

C.7 Calculate the calories lost by the water.

Heat (cal) lost by water = mass of water $\times \Delta T \times$ specific heat (1.00 cal/g °C)

This is the same number of calories that melted the ice.

C.8 Calculate the grams of ice that melted by subtracting the initial mass of the cup and water from the final mass of the cup and water after the ice melted.

C.9 Calculate your experimental value for the heat of fusion for ice.

$$\text{Heat of fusion (cal / g)} \ = \ \frac{\text{heat (cal) gained to melt ice}}{\text{grams of ice}}$$

Report Sheet - Lab 9

Date _____   Name _____

Section _____   Team _____

Instructor _____   _____

Pre-Lab Study Questions

1. Why is energy required for the heating or boiling process?

2. When water at 0°C freezes, is heat lost or gained?

A. A Heating Curve for Water   Volume of water: _____mL

A.1
Time (min)	Temperature (°C)	Time (min)	Temperature (°C)
0	_____	_____	_____
1	_____	_____	_____
2	_____	_____	_____
_____	_____	_____	_____
_____	_____	_____	_____
_____	_____	_____	_____
_____	_____	_____	_____
_____	_____	_____	_____
_____	_____	_____	_____

Report Sheet - Lab 9

A.2 Graphing the Heating Curve

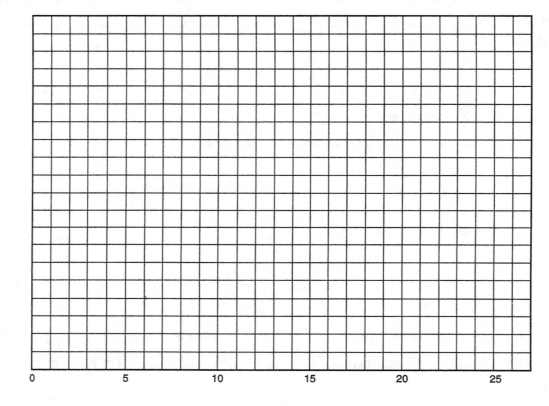

Time (min)

A.3 Boiling point of water   _____°C

A.4 Temperature change (ΔT)   _____°C

A.5 Volume of water   _____ mL

     Mass of water   _____ g

A.6 Number of calories needed to heat water   _____ cal
     (Show calculations.)

Questions and Problems

Q.1 On the heating curve, how long did it take for the temperature to rise to 60°C?

Report Sheet - Lab 9

B. Graphing a Cooling Curve

B.1

Time (min)	Temperature (°C)	Time (min)	Temperature (°C)
0	_____	_____	_____
1	_____	_____	_____
2	_____	_____	_____
3	_____	_____	_____
4	_____	_____	_____
5	_____	_____	_____
_____	_____	_____	_____
_____	_____	_____	_____
_____	_____	_____	_____
_____	_____	_____	_____
_____	_____	_____	_____

B.2 Graphing the Cooling Curve

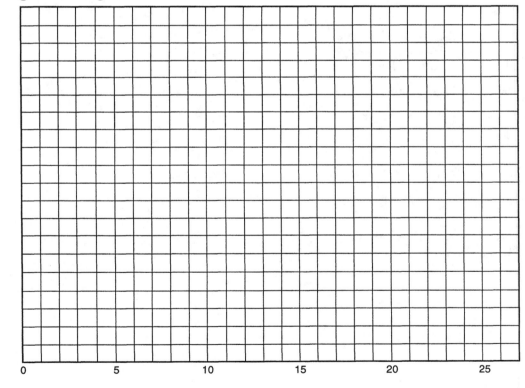

Temperature (C)

Time (min)

Questions and Problems

Q.2 What is the freezing point of Salol?

Report Sheet - Lab 9

C. Energy in Changes of State

C.1 Empty calorimeter cup   _____ g

C.2 Calorimeter + water   _____ g

C.3 *Final* water temperature   _____ °C

    Initial water temperature   _____ °C

C.4 Calorimeter + water + melted ice   _____ g

Calculations

C.5 Mass of water   _____ g

C.6 Temperature change   _____ °C

C.7 Calories lost by water
    (Show calculations.)   _____ cal

    Calories used to melt ice   _____ cal

C.8 Mass of ice that melted
    (Show calculations.)   _____ g

C.9 Heat of fusion (calories to melt 1 g of ice)   _____ cal/g
    (Show calculations.)

Report Sheet - Lab 9

Questions and Problems

Q.3 When water is heated, the temperature eventually reaches a constant value and forms a plateau on the graph. What does the plateau indicate?

Q.4 175 g of water was heated from 15° to 88°C. How many kilocalories were absorbed by the water?

Q.5 How many calories are required at 0°C to melt an ice cube with a mass of 25 g?

Q.6 a. Calculate the amount of heat (kcal) released when 50.0 g of water at 100°C hits the skin and cools to a body temperature of 37°C.

b. Calculate the amount of heat (kcal) released when 50.0 g of steam at 100°C hits the skin, condenses, and cools to a body temperature of 37°C.

c. Use your answer in 6a and 6b to explain why steam burns are so severe.

Chemical Reactions and Equations

Goals

- Observe physical and chemical properties associated with chemical changes.
- Give evidence for the occurrence of a chemical reaction.
- Write a balanced equation for a chemical reaction.
- Identify a reaction as a combination, decomposition, replacement, or combustion reaction.

Discussion

When a substance undergoes a physical change, it changes its appearance but not its composition. For example, when silver (Ag) melts and forms liquid silver (Ag), it undergoes a physical change from solid to liquid. In a chemical change, a substance is changed to give a new substance with a different composition and different properties. For example, when silver tarnishes, the shiny silver (Ag) changes to a dull-gray silver sulfide (Ag_2S), a new substance with different properties and a different composition. See Table 10.1.

Table 10.1 *Comparison of Physical and Chemical Changes*

Some Physical Changes	Some Chemical Changes
Change in state	Formation of a gas (bubbles)
Change in size	Formation of a solid (precipitates)
Tearing	Disappearance of a solid (dissolves)
Breaking	Change in color
Grinding	Heat is given off or absorbed

Balancing a Chemical Equation

In a chemical reaction, atoms in the reactants are rearranged to produce new combinations of atoms in the products. However, the total number of atoms of each element in the reactants is equal to the total number of atoms in the products. In an equation, the reactants are shown on the left and the products on the right. An arrow between them indicates that a chemical reaction takes place.

$$\text{Reactants} \longrightarrow \text{Products}$$

To balance the number of atoms of each element on the left and right sides of the arrow, we write a number called a *coefficient* in front of the formula containing that particular element. Consider the balancing of the following unbalanced equation. The state of the substances as gas is shown as (g).

$N_2(g) + H_2(g) \longrightarrow NH_3(g)$ *Unbalanced equation*

$N_2(g) + H_2(g) \longrightarrow 2NH_3(g)$ *A coefficient of 2 balances the N atoms.*

$N_2(g) + 3H_2(g) \longrightarrow 2NH_3(g)$ *A coefficient of 3 balances the H atoms.*

                                                The equation is now balanced.

Types of Reactions

There are many different chemical reactions, but most can be classified into the types of reactions shown in Table 10.2.

Table 10.2 *Common Types of Chemical Reactions*

Type of Reaction	Description	Example Equation
Combination	Elements or simple compounds form a more complex product.	$Cu + S \rightarrow CuS$
Decomposition	A reacting substance is split into simpler products.	$CaCO_3 \rightarrow CaO + CO_2$
Single replacement	One element takes the place of another element in a compound.	$Mg + 2HCl \rightarrow MgCl_2 + H_2$
Double replacement	Elements in two compounds switch places.	$AgNO_3 + NaCl \rightarrow AgCl + NaNO_3$
Combustion	Reactant and oxygen form an oxide product.	$S + O_2 \rightarrow SO_2$

Lab Information

Time: 2–2$^1/_2$ hr
Comments: Read all the directions and safety instructions carefully.
                Match the labels on bottles and containers with the names of the substances you need.
                Label your containers with the formulas of the chemicals you place in them.
                Be sure that long hair is tied back.
                A Bunsen burner is a potential hazard. Keep your work area clear of books, papers, backpacks, and other potentially flammable items.
                Tear out the report sheets and place them beside the matching procedures.
Related Topics: Chemical change, chemical equation, balancing chemical equations

Experimental Procedures

 **GOGGLES REQUIRED!**

A. Magnesium and Oxygen

Materials: Magnesium ribbon (2–3 cm long), tongs, Bunsen burner

A.1   Obtain a small strip (2–3 cm) of magnesium ribbon. Record its appearance. Using a pair of tongs to hold the end of the magnesium ribbon, ignite it using the flame of a Bunsen burner. *As soon as the magnesium ribbon ignites, remove it from the flame. Shield your eyes as the ribbon burns.* Record your observations of the reaction and the physical properties of the product. Use complete sentences to describe your observations.

A.2   Balance the equation given for the reaction. Use 1 as a coefficient when one unit of that substance is required. The letters in parentheses indicate the physical state of the reactant or product: (*g*) gas, (*s*) solid. *Unbalanced equation:* $Mg(s) + O_2(g) \longrightarrow MgO(s)$

A.3   Identify the type of reaction that has occurred. For this reaction, more than one reaction type may be used to classify the reaction.

B. Zinc and Copper(II) Sulfate

Materials: Two test tubes, test tube rack, 1 M $CuSO_4$ (copper(II) sulfate solution), $Zn(s)$

For all the experiments in parts C–F, use small quantities. For solids, use the amount of compound that will fit on the tip of a spatula or small scoop. Carefully pour small amounts of liquids into your own beakers and other containers. Measure out 3 mL of water in a test tube. Use this volume as a reference level for each of the experiments.

Do not place droppers or stirring rods into reagent bottles. They may contaminate a reagent for the entire class. Discard unused chemicals as indicated by your instructor.

B.1 Pour 3 mL (match the reference volume) of the *blue solution,* 1 M $CuSO_4$ (one molar copper(II) sulfate), into each of two test tubes. Obtain a small piece of zinc metal. Describe the appearance of the $CuSO_4$ solution and the small piece of zinc metal. Add the Zn metal piece to the $CuSO_4$ solution in one of the test tubes. The $CuSO_4$ solution in the other test tube is your reference for the initial solution color. Place the test tubes in your test tube rack and observe the color of the $CuSO_4$ solutions and the Zn piece again at 15 and 30 minutes. Pour the $CuSO_4$ solutions into the sink followed by a large amount of water. Rinse the piece of zinc with water and place it in a recycling container as directed by your instructor.

B.2 Balance the equation given for the reaction. The symbol (*aq*) means aqueous (dissolved in water).
Unbalanced equation: $Zn(s) + CuSO_4(aq) \longrightarrow Cu(s) + ZnSO_4(aq)$

B.3 Identify the type of reaction that has occurred.

C. Metals and HCl

Materials: Three test tubes, test tube rack, small pieces of $Cu(s)$, $Zn(s)$, and $Mg(s)$ metal
1 M HCl ***Caution: HCl is a corrosive acid. Handle carefully!***

C.1 Place 3 mL of 1 M HCl (match your reference volume from part C) in each of three test tubes. Describe the appearance of each metal. Carefully add a metal piece to the acid in each of the test tubes. Record any evidence of reaction such as bubbles of gas (H_2). Carefully pour off the acid and follow with large quantities of water to dilute. Rinse the metal pieces with water, dry, and return to your instructor.

C.2 Balance the equation given for each metal that gave a chemical reaction. If there was no reaction, cross out the products and write NR for no reaction.

Unbalanced equations: 1. $Cu(s) + HCl(aq) \longrightarrow CuCl_2(aq) + H_2(g)$

2. $Zn(s) + HCl(aq) \longrightarrow ZnCl_2(aq) + H_2(g)$

3. $Mg(s) + HCl(aq) \longrightarrow MgCl_2(aq) + H_2(g)$

C.3 Identify the type of reaction for each chemical reaction that occurred.

D. Reactions of Ionic Compounds

Materials: Three (3) test tubes, test tube rack
Dropper bottle sets of 0.1 M solutions: $CaCl_2$, Na_3PO_4, $BaCl_2$, Na_2SO_4, $FeCl_3$, KSCN

For each of these reactions, two substances will be mixed together. Describe your observations of the reactants before you mix them and then describe the products of the reaction. Look for changes in color, the formation of a solid (solution turns cloudy), the dissolving of a solid, and/or the formation of a gas (bubbling). Balance the equations for the reactions. Dispose of the solutions properly.

D.1 Place 20 drops each of 0.1 M $CaCl_2$ (calcium chloride) and 0.1 M Na_3PO_4 (sodium phosphate) into a test tube. Describe any changes that occur. Identify the type of reaction for each chemical reaction that occurred. *Unbalanced equation:* $CaCl_2(aq) + Na_3PO_4(aq) \longrightarrow Ca_3(PO_4)_2(s) + NaCl(aq)$

D.2 Place 20 drops each of 0.1 M $BaCl_2$ (barium chloride) and 0.1 M Na_2SO_4 (sodium sulfate) into a test tube. Describe any changes that occur. Identify the type of reaction for each chemical reaction that occurred. *Unbalanced equation:* $BaCl_2(aq) + Na_2SO_4(aq) \longrightarrow BaSO_4(s) + NaCl(aq)$

D.3 Place 20 drops each of 0.1 M $FeCl_3$ (iron(III) chloride) and 0.1 M KSCN (potassium thiocyanate) into a test tube. Describe any changes that occur. Identify the type of reaction for each chemical reaction that occurred. *Unbalanced equation:*
$$FeCl_3(aq) + KSCN(aq \longrightarrow Fe(SCN)_3(aq) + KCl(aq)$$

E. Sodium Carbonate and HCl

Materials: Test tube, test tube rack, 1 M HCl solution, $Na_2CO_3(s)$, and matches or wood splints

E.1 Place about 3 mL of 1 M HCl in a test tube. Add a small amount of solid Na_2CO_3 (about the size of a pea) to the test tube. Record your observations. ***Caution: HCl is corrosive. Clean up any spills immediately. If spilled on the skin, flood the area with water for at least 10 minutes.***

E.2 Identify the type of reaction for each chemical reaction that occurred.
Unbalanced equation: $Na_2CO_3(s) + HCl(aq) \longrightarrow CO_2(g) + H_2O(l) + NaCl(aq)$

E.3 Light a match or wood splint and insert the flame inside the neck of the test tube. What happens to the flame? Record your observations.

Report Sheet - Lab 10

Date _____   Name _____

Section _____   Team _____

Instructor _____   _____

Pre-Lab Study Questions

1. Why is the freezing of water called a physical change?

2. Why are burning candles and rusting nails examples of chemical change?

3. What is included in a chemical equation?

4. How does a combination reaction differ from a decomposition reaction?

A.   Magnesium and Oxygen

A.1 Initial appearance of Mg   _____

                                       _____

      Observations of the reaction   _____

                                       _____

      Appearance of the product   _____

                                       _____

A.2 Balance: _____Mg(s)   +   _____ O$_2$(g) $\longrightarrow$ _____ MgO(s)

A.3 Type of reaction: _____

                                       _____

Report Sheet - Lab 10

B. Zinc and Copper(II) Sulfate

B.1 Initially   Zn   _____

             $CuSO_4$ _____

   15 min   Zn   _____

             $CuSO_4$ _____

   30 min   Zn   _____

             $CuSO_4$ _____

B.2   Balance: ____ $Zn(s)$ + ____ $CuSO_4(aq)$ $\longrightarrow$ ____ $Cu(s)$ + ____ $ZnSO_4(aq)$

B.3   Type of reaction: _____

_____

C. Metals and HCl

C.1   Observations

   Cu   Initial: _____

             Reaction: _____

   Zn   Initial: _____

             Reaction: _____

   Mg   Initial: _____

             Reaction: _____

C.2   Balance: ____$Cu(s)$ + ____ $HCl(aq)$ $\longrightarrow$ ____$CuCl_2(aq)$ + ____ $H_2(g)$

             ____$Zn(s)$ + ____ $HCl(aq)$ $\longrightarrow$ ____$ZnCl_2(aq)$ + ____ $H_2(g)$

             ____$Mg(s)$ + ____ $HCl(aq)$ $\longrightarrow$ ____$MgCl_2(aq)$ + ____ $H_2(g)$

C.3   Type of reaction: Cu _____

                       Zn _____

                       Mg _____

Report Sheet - Lab 10

D. Reactions of Ionic Compounds

D.1 **CaCl₂ and Na₃PO₄**

Observations:_____

_____

Type of reaction: _____

Balance: ____$CaCl_2(aq)$ + ____ $Na_3PO_4(aq)$ $\longrightarrow$ ____ $Ca_3(PO_4)_2(s)$ + ____ $NaCl(aq)$

D.2 **BaCl₂ and Na₂SO₄**

Observations:_____

_____

Type of reaction: _____

Balance: ____ $BaCl_2(aq)$ + ____ $Na_2SO_4(aq)$ $\longrightarrow$ ____ $BaSO_4(s)$ + ____ $NaCl(aq)$

D.3 **FeCl₃ and KSCN**

Observations:_____

_____

Type of reaction: _____

Balance: ____ $FeCl_3(aq)$ + ____ $KSCN(aq)$ $\longrightarrow$ ____ $Fe(SCN)_3(aq)$ + ____ $KCl(aq)$

E. Sodium Carbonate and HCl

E.1 Observations:_____

_____

E.2 Type of reaction: _____

Balance: ____ $Na_2CO_3(s)$ + ____ $HCl(aq)$ $\longrightarrow$ ____ $CO_2(g)$ + ____ $H_2O(l)$ + ____ $NaCl(aq)$

E.3 Why did the flame of the burning match or splint go out?

Report Sheet - Lab 10

Questions and Problems

Q.1 What evidence of a chemical reaction might you see in the following cases?

a. Dropping an Alka-Seltzer™ tablet into a glass of water

_____

b. Bleaching a stain

_____

c. Burning a match

_____

d. Rusting of an iron nail

_____

Q.2 Balance the following equations:

a. _____ $Mg(s)$ + _____ $HCl(aq)$ $\longrightarrow$ _____ $H_2(g)$ + _____ $MgCl_2(aq)$

b. _____ $Al(s)$ + _____ $O_2(g)$ $\longrightarrow$ _____ $Al_2O_3(s)$

c. _____ $Fe_2O_3(s)$ + _____ $H_2O(l)$ $\longrightarrow$ _____ $Fe(OH)_3(s)$

d. _____ $Ca(OH)_2(aq)$ + _____ $HNO_3(aq)$ $\longrightarrow$ _____ $Ca(NO_3)_2(aq)$ + _____ $H_2O(l)$

Q.3 Write an equation for the following reactions. Remember that gases of elements such as oxygen are diatomic (O_2). Write the *correct formulas* of the reactants and products. Then correctly balance each equation.

a. Potassium and oxygen gas react to form potassium oxide.

b. Sodium and water react to form sodium hydroxide and hydrogen gas.

c. Iron and oxygen gas react to form iron(III) oxide.

Report Sheet - Lab 10

Q.4 Classify each reaction as combination (C), decomposition (DC), single replacement (SR), or double replacement (DR).

a. $Ni + F_2 \longrightarrow NiF_2$ _____

b. $Fe_2O_3 + 3C \longrightarrow 2Fe + 3CO$ _____

c. $CaCO_3 \longrightarrow CaO + CO_2$ _____

d. $H_2SO_4 + 2KOH \longrightarrow K_2SO_4 + 2H_2O$ _____

Q.5 Predict what product(s) would form from the reaction of the following reactants:

a. $Zn + CuBr_2 \longrightarrow$ _____ + _____

b. $H_2 + Cl_2 \longrightarrow$ _____

c. $MgCO_3 \longrightarrow$ _____ + _____

d. $KCl + AgNO_3 \longrightarrow$ _____ + _____

Reaction Rates and Equilibrium

Goals

- Identify a reaction as exothermic or endothermic.
- Identify the factors that affect the rate of a reaction.
- Observe that chemical reactions are reversible.
- Discuss factors that cause a shift in equilibrium.

Discussion

A. Exothermic and Endothermic Reactions

In an *exothermic* reaction, heat is released, which causes the temperature of the surroundings to increase. An *endothermic* reaction absorbs heat, which causes a drop in the temperature of the surroundings. Heat can be written as a product in an equation for an exothermic reaction and as a reactant in the equation for an endothermic reaction. Energy is required to break apart bonds and is released when bonds form. If energy is released by forming bonds, the reaction is exothermic. If energy is required to break apart bonds, the reaction is endothermic. In our cells, the bonds in carbohydrates are broken down to give us energy. Reactions that build molecules and repair cells are endothermic because they require energy.

Exothermic reactions: $C + O_2 \longrightarrow CO_2 + heat$

$C_6H_{12}O_6 + 6O_2 \longrightarrow 6CO_2 + 6H_2O + energy$
Glucose

Endothermic reactions: $Heat + PCl_5 \longrightarrow PCl_3 + Cl_2$

$Energy + amino\ acids \longrightarrow protein$

B. Rates of Reactions

The rate or speed at which a reaction occurs depends on the *amounts of the reactants*, the temperature, and the presence of a *catalyst*.

Amount of Reactants If more reactant is added, products form faster. For example, we normally breathe air that is 20% oxygen (O_2). However, if a person is given pure oxygen (100%), oxygenated hemoglobin (HbO_2) is formed faster.

$$Hb + O_2 \longrightarrow HbO_2$$

If some reactant is removed, the rate at which product forms is slowed. At higher altitudes the ability of hemoglobin to pick up O_2 is slowed because there is a lower amount of O_2. This results in less O_2 reaching the cells in the body and especially the brain. A lowered level of O_2 in the brain may result in mental confusion, hallucinations, and poor decisions, which has led to disastrous results for people who have tried to climb Mt. Everest and other high mountains without adequate sources of oxygen.

Temperature Typically a reaction goes at a faster rate as the temperature is increased. In general, the rate of a reaction doubles for a 10°C increase. We heat food to make it cook faster. When we have a fever, our metabolic reactions including our rate of breathing and our pulse go faster. In cardiac surgery, the body temperature is lowered to slow down the metabolic reactions and the amount of oxygen required by the brain.

In a chemical reaction, collisions between the reactants can lead to the formation of products. The energy that is needed to change the reactants into products is called the *energy of activation*. At high temperatures, more collisions have the energy to react and form products. At low temperature, few collisions lead to products.

Catalyst A typical catalyst increases the rate of a reaction without becoming a part of the product. Enzymes are biological catalysts that make components in our cells react at the rates required for cellular survival. Industry also makes use of catalysts. For example, the reaction of hydrogen with vegetable oils to produce margarine goes faster with a platinum (Pt) catalyst.

$$\text{Vegetable oil} + H_2 \xrightarrow{\text{Pt}} \text{margarine}$$

When a catalyst is present, the reaction can take an alternative pathway, which has a lower energy of activation. Then more reactants have the energy upon collision to change to products.

C. Reversible Reactions

When a reaction begins, reactants are converted to products. That means that the number of product atoms or molecules will increase and collisions will occur between the products. In some of the product collisions, bonds are broken and reactant particles re-formed. Thus, most reactions proceed in two directions—forward (reactants to products) and reverse (products back to reactants). In that case, the reaction is called a *reversible* reaction.

Eventually, the rate of the forward reaction becomes equal to the rate of the reverse reaction. There is no further change in the amounts of reactants and products and the system is at equilibrium. Some reversible reactions reach equilibrium quickly, but others may take a very long time. However, in any equilibrium, all of the reactants and products are present. For example, the reaction of ammonia (NH_3) and water produces ammonium ion (NH_4^+) and hydroxide ion (OH^-). At equilibrium, the products are always present along with the reactant molecules.

$$NH_3 + H_2O \rightleftharpoons NH_4^+ + OH^-$$

After this reversible reaction has reached equilibrium, any changes in the amounts of one of the reactants or products will create a stress. An increase in the amount of a reactant will favor the forward reaction. If a product is added, the reverse reaction will be favored. In a similar way, removal of reactants will shift the equilibrium in the reverse direction. Removing a product will favor the forward reaction.

D. Iron (III)-thiocyanate Equilibrium

In another reversible reaction the yellow Fe^{3+} ion and the colorless SCN^- (thiocyanate) ion, are in equilibrium with a deep red complex ion $FeSCN^{2+}$. More SCN^- forms $Fe(SCN)_2^+$.

$$Fe^{3+} (aq) + SCN^- (aq) \rightleftharpoons FeSCN^{2+}$$
$$\text{yellow} \qquad \text{colorless} \qquad\qquad \text{red}$$

$$Fe^{3+} (aq) + 2\ SCN^- (aq) \rightleftharpoons Fe(SCN)_2^+$$
$$\text{yellow} \qquad \text{colorless} \qquad\qquad \text{red}$$

When the system contains mostly reactants, the solution is yellow. When the system shifts to products, the color changes to the deep red of $FeSCN^{2+}$.

When a system is at equilibrium, the rate of the forward reaction is equal to the rate of the reverse reaction. Then an *equilibrium constant* can be written, which represents the ratio of concentrations of the reactants and products.

$$K_{eq} = \frac{[products]}{[reactants]} = \frac{[FeSCN^{2+}]}{[Fe^{3+}][SCN^-]}$$

Visual colors allow us to see a shift in equilibrium between reactants and products. If we add more Fe^{3+} or SCN^-, the equilibrium shifts to product, which is red. If we add a substance that removes Fe^{3+} or SCN^-, the equilibrium shifts back to reactants. Then we will see the formation of the yellow color of the iron (III) ion.

One way to reduce the yellow color of Fe^{3+} is to add Cl^-, which forms a colorless $FeCl_4^-$ complex.

$$Fe^{3+} + 4\ Cl^- \rightleftharpoons FeCl_4^-$$
$$\text{yellow} \qquad\qquad\qquad \text{colorless}$$

Lab Information

Time: 2– 2 ½ hr

Comments: Tear out the report sheets and place them next to the corresponding procedures. Carefully observe the colors and state of the components in the test tubes before and after reactions.

Related topics: Factors affecting rates of reactions, energy of activation, chemical equilibrium, factors affecting equilibrium

Experimental Procedures

Goggles are required!

A. Exothermic and Endothermic Reactions

Materials: Two test tubes, test tube rack, water, scoop or spatula
$NH_4NO_3(s)$, anhydrous $CaCl_2(s)$, thermometer

A.1 Place 5 mL of water in each of two test tubes. Record the temperature of the water. Add one scoop of $NH_4NO_3(s)$ crystals to the water in the first test tube. Add one scoop of *anhydrous* $CaCl_2(s)$ to the water in the second test tube. *Anhydrous* means "without water." Stir each and record the temperature again.

$$NH_4NO_3(s) \xrightarrow{H_2O} NH_4^+(aq) + NO_3^-(aq)$$

$$CaCl_2(s) \xrightarrow{H_2O} Ca^{2+}(aq) + 2\ Cl^-(aq)$$

A.2 Describe each reaction as endothermic or exothermic.

A.3 To each equation, add the term heat on the side of the reactants (if endothermic) and on the side of the products (if exothermic).

B. Rates of Reactions

Materials: Test tubes, cleaned pieces of magnesium ribbon 2–3 cm long (about 0.4 g)
1.0 M HCl, 2.0 M HCl, 3.0 M HCl
Two 250-mL beakers, 400-mL beaker for hot water bath, ice
Vinegar (or 0.1 M HCl), $NaHCO_3$ (or baking soda or Alka-Seltzer), scoop or spatula, thermometer

B.1 **Amount of reactants** Obtain a piece of Mg and 10 mL of 1.0 M HCl. Place the Mg in a test tube. In this test you are going to measure the time required to completely dissolve the Mg. First note the time on a watch with a second hand or on a stopwatch. Then pour 10 mL of 1.0 M HCl solution into the test tube all at once. Stir. Record the amount of time for the Mg to completely react and disappear. At the end of the reaction determine whether the reaction is exothermic or endothermic by touching the bottom of the test tube.

Caution: HCl is an acid. Work carefully. If any HCl gets in your eyes wash immediately at the eye fountain.

Repeat the reaction of Mg and the other HCl solutions. Alternatively this work could be shared with other students. Record the time for each sample to completely react. Rank the rates of reactions from slowest to fastest.

Be sure to wash your hands at the end of this experiment.

B.2 **Temperature** Place 10 mL of vinegar (or 0.1 M HCl) in each of two large test tubes. Place one test tube in a 250-mL beaker half-filled with crushed ice. Cool to a temperature of 10°C or lower. Place the other test tube and vinegar in a 400-mL beaker about half-filled with water. Heat to 50–60°C. Remove the test tubes and place them in a test tube rack. Measure the temperature in each test tube. At the same time, add 1 scoop or 2 spatula tips of $NaHCO_3$ to each sample. Observe and compare the fizzing (bubbles) in each test tube. Determine which test tube clears first.

$$NaHCO_3 + H^+ \longrightarrow CO_2(g) + Na^+ + H_2O$$

C. Reversible Reactions

Materials: 3 test tubes, test tube rack, and droppers,
0.1 M $CuCl_2$, 0.1 M NaOH, 1 M NH_4OH, 0.1 M HCl or dropper bottle sets

The Cu^{2+} ion reacts with OH^- to form solid $Cu(OH)_2$, which is in equilibrium with the ions according to the following:

$$Cu(OH)_2(s) \rightleftharpoons Cu^{2+}(aq) + 2OH^-(aq)$$

Place 3 mL of 0.1 M $CuCl_2$ in each of three test tubes. To the first test tube add drops of 0.1 M NaOH until a white, cloudy precipitate of $Cu(OH)_2$ starts to form. Add the same number of drops to the other two test tubes containing $CuCl_2$ to give solid $Cu(OH)_2$.

C.1 **Test tube 1** Describe the initial appearance of the test tube. Using a dropper, add more 0.1 M NaOH, which increases the amount of OH^-. Describe the change in the appearance of the sample in the test tube. In the equilibrium equation for $Cu(OH)_2$, identify the component that increased or decreased. Determine how the equilibrium shifted in response to this stress.

C.2 **Test tube 2** Describe the initial appearance of the test tube. Using a dropper, add drops of 1 M NH_4OH ($NH_3 + H_2O$). Describe the change of appearance of the sample in the test tube. NH_3 reacts with CU^{2+} to form the deep blue ion $Cu(NH_3)_4^{2+}$, which decreases the amount of Cu^{2+} in the equilibrium system for $Cu(OH)_2$.

$$Cu^{2+} + 4NH_3 \longrightarrow Cu(NH_3)_4^{2+}$$
deep blue

In the equilibrium equation for $Cu(OH)_2$, identify the component that increased or decreased. Determine how the equilibrium shifted in response to this stress.

C.3 **Test tube 3** Using a dropper, add a few drops of 0.1 M HCl. Describe the change in the appearance of the sample in the test tube. H^+ from HCl reacts with the OH^- in the equilibrium system to form water, which decreases the amount of OH^-.

$$H^+ + OH^- \longrightarrow H_2O$$

In the equilibrium equation for $Cu(OH)_2$, identify the component that increased or decreased. Determine how the equilibrium shifted in response to this stress.

D. Iron(III)-thiocyanate Equilibrium

Materials: 6 test tubes, test tube rack, small 100-mL beaker, 0.01 M $Fe(NO_3)_3$, 3 cork stoppers, 1 M $Fe(NO_3)_3$, 0.01 M KSCN, 1 M KSCN, 3 M HCl, 3 M NaOH, water, two 250-mL beakers, Bunsen burner, ice

Using a small beaker, prepare a stock equilibrium solution by mixing 10 mL of 0.01 M $Fe(NO_3)_3$ and 10 mL of 0.01 M KSCN.

D.1 Set up 6 test tubes in a test tube rack and label each. To each test tube, add 3 mL of the stock equilibrium solution. To tube 1 add 10 drops of water. The first test tube will be your control, which means that you compare its color to the colors that form in the other test tubes. No other reagents will be added to this control test tube. Record the color you observe in the control tube 1.

D.2 To tube 2, add slowly 10 drops of 1 M $Fe(NO_3)_3$. Place a cork stopper in the test tube and invert it several times to mix the contents. Record the color you observe in tube 2 and compare to tube 1.

D.3 To tube 3, add slowly 10 drops of 1 M KSCN. Place a cork stopper in the test tube and mix as before. Record the color you observe in tube 3 and any change in color.

D.4 To tube 4, add slowly 10 drops of 3 M HCl. Place a cork stopper in the test tube and mix as before. Record the color you observe in tube 4 and any change in color.

For the following, prepare a hot water bath by filling a 250-mL beaker about one half full of water and heating. Use a second 250-mL beaker as an ice bath by filling it about one half full with ice and water.

D.5 To tube 5, add slowly 10 drops of water and place the test tube in a beaker of warm water. Turn off the heat source to prevent the water from boiling. After 10 minutes, record the color you observe in tube 5 and any change in color.

D.6 To tube 6, add slowly 10 drops of water and place the test tube in a beaker of ice. After 10 minutes, record the color you observe in tube 6 and any change in color.

Report Sheet - Lab 11

Date _____   Name _____

Section _____   Team _____

Instructor _____

Pre-Lab Study Questions

1. How does an exothermic reaction differ from an endothermic reaction?

2. What factors increase the rate of a chemical reaction?

3. When is equilibrium established in a reversible reaction?

4. How does a system at equilibrium respond to the addition of more reactant? More product?

A. Exothermic and Endothermic Reactions

A.1 NH_4NO_3 $CaCl_2$

  Initial temperature   _____   _____

  Final temperature   _____   _____

  Temperature change   _____   _____

A.2   Endothermic or exothermic   _____   _____

A.3   Equations (add *heat*)

$$NH_4NO_3(s) \xrightarrow{H_2O} NH_4^+(aq) + NO_3^-(aq)$$

$$CaCl_2(s) \xrightarrow{H_2O} Ca^{2+}(aq) + 2\ Cl^-(aq)$$

Report Sheet - Lab 11

Questions and Problems

Q.1 As a lab technician for a pharmaceutical company, you are responsible for preparing hot packs and cold packs. A hot pack involves the release of heat when a salt and water are mixed. A cold pack becomes colder because mixing a salt and water absorbs heat.

Which compound could you use to make a hot pack?

Which compound could you use to make a cold pack?

Q.2 When you burn a log in the fireplace or burn gasoline in a car, is the reaction (combustion) an endothermic or exothermic reaction?

B. Rates of Reactions

B.1 Amount of reactants

	Time (sec) for complete reaction	Exothermic or endothermic	Reaction rate from slowest to fastest
1.0 M HCl			
2.0 M HCl			
3.0 M HCl			

Questions and Problems

Q.3 How does the amount of reactant account for the differences in the rate of reaction?

Report Sheet - Lab 11

B.2 **Test Tube**   **Temperature**   **Observations**

1   _____   _____

_____

2   _____   _____

_____

Which test tube cleared first? _____

Questions and Problems

Q.4 How does temperature affect the rate of a reaction?

C. Reversible Reactions

$$Cu(OH)_2(s) \rightleftharpoons Cu^{2+}(aq) + 2OH^-(aq)$$

Initial color of the $CuCl_2$ solution   _____

Color and appearance of $Cu(OH)_2(s)$ _____

	C.1 Test tube 1	C.2 Test tube 2	C.3 Test tube 3
Initial appearance			
Change in appearance			
What component in the equilibrium equation increased or decreased?			
Did the equilibrium shift to reactants or products?			

Report Sheet - Lab 11

Questions and Problems

Q.5 What is meant by the term *reversible* reaction?

Q.6 Explain how a change in the amount of a product causes a shift in equilibrium.

Q.7 Predict the direction that equilibrium will shift for each change in the components of the following reaction:

$$C(s) + H_2O(g) + heat \rightleftharpoons CO(g) + H_2(g)$$

a. Adding heat

b. Adding $CO(g)$

c. Removing $H_2(g)$

d. Adding $H_2O(g)$

Q.8 Write the equilibrium expression (K_{eq}) for the reaction:

$$PCl_5(g) \rightleftharpoons PCl_3(g) + Cl_2(g)$$

Report Sheet - Lab 11

D.   Iron(III)-thiocyanate Equilibrium

	Test tube	Initial color	Final color	Ions that increase	Ions that decrease	Equilibrium shifted toward
D.1	1					
D.2	2					
D.3	3					
D.4	4					
D.5	5					
D.6	6					

Questions and Problems

Q. 9 Indicate whether the addition of each of the following increases or decreases the $FeSCN^{2+}$ in an equilibrium mixture. Explain.

Add	Increases $FeSCN^{2+}$	Decreases $FeSCN^{2+}$	Reason
a. Fe^{3+}			
b. Heat			
c. Cl^-			
d. Cooling			
e. SCN^-			

Moles and Chemical Formulas

Goals

- Use the mole conversion factors to convert grams to moles and moles to grams.
- Experimentally determine the simplest formula of an oxide of magnesium.
- Calculate the percent water in a hydrated salt.
- Determine the formula of a hydrate.

Discussion

A.   Finding the Simplest Formula

In this experiment, magnesium metal is heated to a high temperature until it reacts with the oxygen (O_2) in the air.

Reactants			Heat	**Product**
$2Mg(s)$	+	$O_2(g)$	$\longrightarrow$	$2MgO(s)$
silvery metal				*white-gray ash*

As the magnesium burns in the air, it may also combine with the nitrogen (N_2) in the air. To remove any nitride product, water is added and the product is reheated. Any nitride product is converted to magnesium oxide and ammonia.

$$3Mg(s) \quad + \quad N_2(g) \quad \xrightarrow{\text{Heat}} \quad Mg_3N_2(s)$$

$$Mg_3N_2(s) \quad + \quad 3H_2O(l) \quad \xrightarrow{\text{Heat}} \quad 3MgO(s) \quad + \quad 2NH_3(g)$$

At the beginning of the experiment, you will obtain the mass of the magnesium metal. At the end, you will determine the mass of the oxide product. The mass of oxygen that combined with the magnesium is the difference between the mass of the oxide product and the original mass of magnesium.

g oxide product   –   g Mg   =   g O that combined to form the oxide product

The simplest formula of the magnesium oxide product is determined by calculating the moles of magnesium and the moles of oxygen using their respective molar masses.

$$\text{moles Mg} = \text{grams of Mg ribbon} \times \frac{1 \text{ mole Mg}}{24.3 \text{ g Mg}}$$

$$\text{moles O} = \text{grams of O combined} \times \frac{1 \text{ mole O}}{16.0 \text{ g O}}$$

The simplest formula is obtained by dividing the number of moles of the elements by the smaller number of moles. Suppose that in a similar experiment we determined that 0.040 mole of Zn had combined with 0.080 mole of Cl to form a compound. We can proceed as follows:

1. Ratio of moles: 0.040 mole Zn : 0.080 mole Cl.

2. Determine the smaller number of moles: 0.040 mole Zn.

3. Divide the moles of each element by the smaller number of moles and round to the nearest whole number.

$$\frac{0.080 \text{ mole Cl}}{0.040} = 2 \text{ moles Cl} \qquad \frac{0.040 \text{ mole Zn}}{0.040} = 1 \text{ mole Zn}$$

4. Use the whole numbers as subscripts to write the formula of the compound.
Zn_1Cl_2 or $ZnCl_2$

B. Formula of a Hydrate

A *hydrate* is a salt that contains a specific number of water molecules called the *water of hydration*. The number of water molecules is fixed for each kind of hydrate, but differs from one salt to another. In the formula of a hydrate, the number of water molecules is written after the salt formula and is separated by a large, raised dot.

$CaSO_4 \bullet H_2O$ $CuSO_4 \bullet 5H_2O$ $Na_2CO_3 \bullet 10H_2O$

Heating the hydrate provides the energy to remove the water molecules. The salt without water is called an *anhydrate*.

$$CuSO_4 \bullet 5H_2O \xrightarrow{\text{Heat}} CuSO_4 + 5 H_2O \text{ (g)}$$
Hydrate   anhydrate   water of hydration

Lab Information

Time: 2–2½ hr
Comments: Tear out the report sheets and place them beside the matching procedures.
                 Use steel wool to remove any dull coating on the magnesium ribbon until it is shiny.
                 Check the crucible for cracks before you start to heat it.
                 When you set a hot object aside to cool, remember that it is hot.
Related Topics: Formulas, moles, molar mass, calculating moles from grams, calculating grams from moles, hydrates, dehydration

Experimental Procedures

 **GOGGLES REQUIRED!**

A. Finding the Simplest Formula

Materials: Crucible, crucible cover, crucible tongs, clay triangle, iron ring, ring stand, Bunsen burner, magnesium ribbon, steel wool, eyedropper, small 100- or 150-mL beaker, laboratory balance, heat-resistant pad

A.1 Obtain a clean, dry crucible and its cover. (A porcelain crucible may have stains in the porcelain that cannot be removed.) Place the crucible and cover in a clay triangle that is sitting on an iron ring attached to a ring stand. See Figure 12.1. Heat for 1 minute. Cool (5–10 minutes) until the crucible is at room temperature. Using crucible tongs, carry the crucible and cover to the balance. Weigh the crucible and cover (together) and record. Give all the figures in the mass; do not round off. ***Do not place hot objects on a balance pan.***

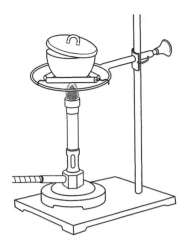

Figure 12.1 Heating a crucible and cover in a clay triangle

A.2   Obtain a piece of magnesium ribbon that has a mass of 0.15–0.3 g. If there is tarnish on the ribbon, remove it by polishing the ribbon with steel wool. Describe the appearance of the magnesium ribbon. Wind the ribbon into a coil and place the coil into a crucible. Weigh the crucible and cover with the magnesium. Record the total mass. *Do not round off mass.*

Do this part of the experiment in a fume hood: Place the crucible and magnesium ribbon in a clay triangle. Have the cover and a pair of tongs on the desktop. Begin to heat the crucible. Watch closely for smoke or fumes, which is the first indication of the magnesium and oxygen reacting. When the magnesium bursts into flame, use tongs to replace the cover on the crucible. If the magnesium does not burn, check the flame. The tip of the inner flame must touch the bottom of the crucible. The bottom of the crucible will be red hot. *Caution: As soon as the magnesium begins to smoke or bursts into flame, place the cover over the crucible using your tongs. Avoid looking directly at the bright flame of the burning magnesium.*

Observe the reaction of the magnesium and oxygen about every minute. When the magnesium *no longer* produces smoke or a flame, remove the cover and place it on a heat-resistant surface. Heat the crucible strongly for another 5 minutes. Turn off the burner and let the crucible and its contents cool to room temperature. The crucible can be moved to the heat-resistant surface using crucible tongs. *Caution: The crucible and iron ring are still hot.*

During the heating, some of the magnesium reacts with nitrogen in the air to form a magnesium nitride product. To remove this nitride product, *carefully* add 15–20 drops of water to the *cooled* contents. Heat the covered crucible and its contents gently for 5 minutes to evaporate the water (some water may splatter). Then heat strongly for 5 minutes. *Caution: Avoid breathing fumes from the crucible because ammonia may be released during the heating.*

A.3   Allow the crucible to cool. You may wish to move the crucible and cover to a heat-resistant surface where it can cool completely. Reweigh the crucible, cover, and oxide contents. Record their total mass. Describe the appearance of the oxide product.

Calculations

A.4   Determine the mass of the magnesium ribbon present originally.

A.5   Calculate the mass of the magnesium oxide product.

A.6   Calculate the mass of oxygen that combined with the magnesium by subtracting the mass of Mg originally present from the final mass of the MgO product.

g MgO product – g Mg = g O (reactant)

A.7 Determine the number of moles of magnesium present in the ash product by dividing the mass of the magnesium by its molar mass.

$$g\ Mg\ \times\ \frac{1\ mole\ Mg}{24.3\ g\ Mg}$$

A.8 Determine the number of moles of oxygen present in the ash product by dividing the mass of the oxygen by its molar mass.

$$g\ O\ \times\ \frac{1\ mole\ O}{16.0\ g\ O}$$

A.9 In their chemical formula, the moles of magnesium in proportion to the moles of oxygen are stated as small whole numbers. Now we need to change the calculated moles of magnesium and oxygen to whole numbers. To do this, divide both values by the smaller one. That makes the smaller value 1. If our lab work and weighing have been carefully done, the other number should be very close to another whole number such as 1, 2, … . Round the results to the nearest whole number.

A.10 Use the whole number values obtained in A.9 as subscripts in writing the simplest formula of the product of magnesium and oxygen.

B. Formula of a Hydrate

Materials: Crucible, clay triangle, crucible tongs, hydrate of $MgSO_4$, iron ring and stand, Bunsen burner, heat-resistant pad, laboratory balance

B.1 Obtain a clean, dry crucible, heat it for 2–3 minutes, and let it cool. Weigh carefully.

B.2 Fill the crucible about 1/3 full with a hydrate of $MgSO_4$. Record the total mass of the crucible and the hydrate.

B.3 Set the crucible and hydrate on a clay triangle that is set on an iron ring. See Figure 12.1. Heat gently for 5 minutes. Then increase the intensity of the flame and heat strongly for 10 minutes more. The bottom of the crucible should become a dull red color. Turn off the burner. Using crucible tongs, move the crucible to a heat-resistant surface. After the crucible cools to room temperature, weigh the crucible and its contents. Record the mass. ***Caution: Allow heated items to cool to room temperature. Do not place a hot container on a balance pan.***

Optional To be sure that you have completely driven off the water of hydration, heat the crucible and its contents for another 5 minutes and cool. Reweigh. If the mass in the second heating is within 0.05 g of the mass obtained after the first heating, you have completely dehydrated the salt. If not, heat again until you have agreement between final masses. Use the *final mass* for your calculations.

Calculations

B.4 Calculate the mass of the hydrate present.

B.5 Calculate the mass of the dry product (anhydrate) after heating.

B.6 Calculate the mass of water released from the hydrate sample by subtracting the mass of the anhydrate from the original mass of the hydrate.

g salt hydrate – g salt anhydrate = g H_2O

B.7 Calculate the percent H_2O in the hydrate by dividing the mass of H_2O by the original mass of the hydrate.

$$\frac{g\ H_2O}{g\ \text{hydrate}} \times 100 = \%\ H_2O\ \text{in hydrate}$$

B.8 Calculate the moles of H_2O in the hydrate by dividing the mass of the H_2O by its molar mass.

$$g\,H_2O \times \frac{1\ \text{mole}}{18.0\ g\,H_2O} = \text{mole}\ H_2O$$

B.9 Calculate the moles of anhydrate by dividing the mass of the anhydrate by its molar mass (120.4 g/mole of $MgSO_4$)

$$g\ \text{anhydrate (see B.5)} \times \frac{1\ \text{mole anhydrate}}{120.4\ g\ \text{anhydrate}} = \text{moles anhydrate}$$

B.10 To determine the formula of the hydrate we need the following ratio:

_____ moles of water to 1 mole of anhydrate

To determine the ratio of moles of water to 1 mole of anhydrate, divide the moles of water (see B.8) by the moles of anhydrate (B.9). Round off the value for moles of H_2O to the nearest whole number.

$$\frac{\text{Moles water (B.8)}}{\text{Moles anhydrate (B.9)}} = \frac{\text{moles}\ H_2O}{1\ \text{mole anhydrate}}$$

B.11 Complete the formula of your hydrate by writing in the number of moles of water.

Report Sheet - Lab 12

Date _____   Name _____

Section _____   Team _____

Instructor _____

Pre-Lab Study Questions

1. What is meant by the simplest formula of a compound?

2. How does a hydrate differ from an anhydrate?

3. What happens when a hydrate is heated?

A.   Finding the Simplest Formula

A.1 Mass of empty crucible + cover   _____ g

A.2 Initial appearance of magnesium   _____

     Mass of crucible + cover + magnesium   _____ g

A.3 Mass of crucible + cover + oxide product   _____ g

     Appearance of oxide product   _____

Calculations

A.4 Mass of magnesium   _____ g Mg

A.5 Mass of magnesium oxide product   _____ g product

A.6 Mass of oxygen in the product   _____ g O

A.7 Moles of Mg   _____ mole Mg
     (Show calculations.)

Report Sheet - Lab 12

A.8 Moles of O   _____ mole O
     (Show calculations.)

A.9 _____ mole Mg : _____ mole O : smaller value _____

$$\frac{\boxed{} \text{ moles Mg}}{\boxed{}} = \text{_____ moles Mg (rounded)}$$

$$\frac{\boxed{} \text{ moles O}}{\boxed{}} = \text{_____ moles O (rounded)}$$

A.10 Formula: **Mg   O**
                 $\square$ $\square$ ⟵——— subscripts

Questions and Problems

Q.1 Using the rules for writing the formulas of ionic compounds, write the ions and the correct formula for magnesium oxide.

Q.2 Write a balanced equation for the reaction of the magnesium and the oxygen.

Q.3 How many grams are needed to obtain 1.5 moles of $MgCO_3$?

Report Sheet - Lab 12

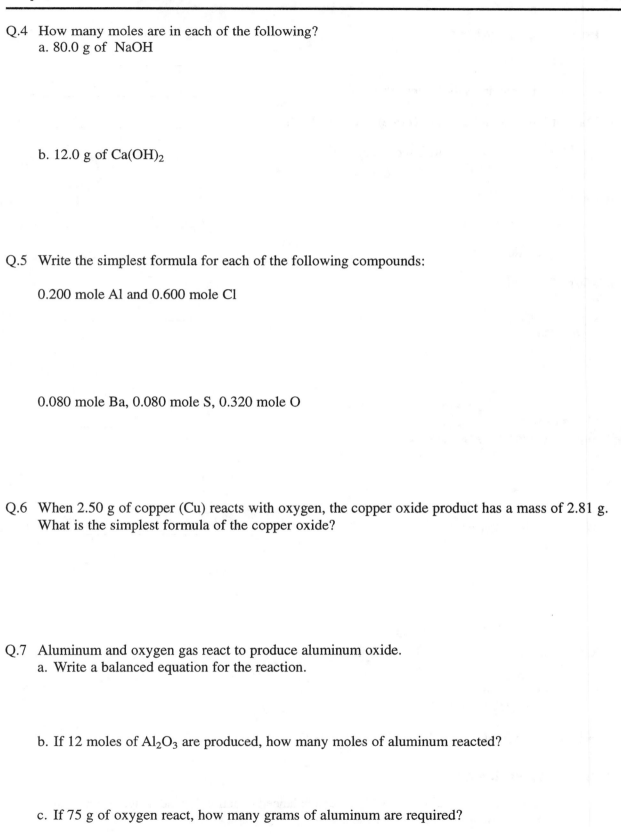

Q.4 How many moles are in each of the following?
   a. 80.0 g of NaOH

   b. 12.0 g of Ca(OH)$_2$

Q.5 Write the simplest formula for each of the following compounds:

   0.200 mole Al and 0.600 mole Cl

   0.080 mole Ba, 0.080 mole S, 0.320 mole O

Q.6 When 2.50 g of copper (Cu) reacts with oxygen, the copper oxide product has a mass of 2.81 g. What is the simplest formula of the copper oxide?

Q.7 Aluminum and oxygen gas react to produce aluminum oxide.
   a. Write a balanced equation for the reaction.

   b. If 12 moles of Al$_2$O$_3$ are produced, how many moles of aluminum reacted?

   c. If 75 g of oxygen react, how many grams of aluminum are required?

Report Sheet - Lab 12

B. Formula of a Hydrate

B.1 Mass of crucible _____ g

B.2 Mass of crucible and salt (hydrate) _____ g

B.3 Mass of crucible and salt (anhydrate) after first heating _____ g

after second heating (optional) _____ g

Calculations

B.4 Mass of salt (hydrate) _____ g

B.5 Mass of salt (anhydrate) _____ g

B.6 Mass of water lost _____ g

B.7 Percent water _____ %
(Show calculations.)

B.8 Moles of water _____ mole
(Show calculations.)

B.9 Moles of salt (anhydrate) _____ mole
(Show calculations.)

B.10 Ratio of moles of water to moles of hydrate

$$\frac{\text{_____ moles water (B.8)}}{\text{_____ moles MgSO}_4 \text{ (B.9)}} = \frac{\text{\_\_\_\_\_ moles H}_2\text{O}}{1 \text{ mole MgSO}_4}$$

B.11 Formula of hydrate (salt) $MgSO_4 \bullet \boxed{} H_2O$

Questions and Problems

Q.8 Using the formula you obtained in B.11, write a balanced equation for the dehydration of the $MgSO_4$ hydrate you used in the experiment.

Gas Laws

Goals

- Graph the relationship between the pressure and volume of a gas.
- Observe the effect of changes in temperature upon the volume of a gas.
- Graph the data for volume and temperature of a gas.
- State a relationship between the temperature and volume of a gas.
- Use a graph to predict the value of absolute zero.

Discussion

Pressure

The pressure of a gas depends upon the number of molecules hitting the walls of a container and their force. You increase pressure when you add air to a car or bicycle tire or basketball or when you blow up a balloon. Airplanes must be pressurized so that you breathe sufficient oxygen. Scuba divers require increased air pressures in their air tanks while diving because the pressure on their bodies increases. In the medical setting, blood pressure is measured to determine the force of the blood against the aorta and other blood vessels. In the operating room, an anesthesiologist monitors the pressures of oxygen and anesthesia given to a patient.

When we work with gases, we obtain the *atmospheric pressure* by reading a barometer in the laboratory. In a barometer, air molecules strike the open surface of mercury, which pushes against the column of mercury in the vertical glass tube. When the height of the mercury column is 760 mm Hg, it measures a pressure of 1 atmosphere (atm). As the atmosphere pressure changes, the height of the mercury in the tube also changes.

A. Boyle's Law

According to Boyle's law, the pressure (P) of a gas varies inversely with the volume (V) of the gas when the temperature (T) and quantity (n) are kept constant. Mathematically, this is expressed as

$$P \propto \frac{1}{V} \ (T, n \text{ constant}) \qquad \text{Boyle's law}$$

By rearranging Boyle's law, we find that pressure × volume of a specific amount of gas is constant as long as the temperature does not change.

$$P_1 V_1 \ = \ \text{constant}$$

At the same temperature, the PV product of a gas remains constant. Therefore, we can write Boyle's law as an equality of the initial (*1*) pressure × volume with the final (2) pressure × volume.

$$P_1 V_1 \ = \ P_2 V_2 \ (T, n \text{ constant})$$

According to Boyle's law, volume decreases when the pressure increases, and volume increases when the pressure decreases. For example, if the pressure of a gas is doubled, the corresponding volume is one-half its previous value.

B. Charles' Law

In cold weather, the tires on a car or a bicycle seem to go flat. In hot weather, a full tire may burst. Such examples of changes in volume are related to changes in temperature. When temperature rises, the kinetic energy of the molecules increases. To keep pressure constant, the volume must expand. When the molecules slow down in cooler weather, the volume must decrease. According to Charles' law, the volume of a gas changes directly with the Kelvin temperature as long as the pressure and number of moles remain constant. Charles' law is expressed mathematically as

$$V \propto T_K \ \ (P \text{ and } n \text{ are constant})$$

$$\text{or } \frac{V}{T_K} = \text{constant}$$

Under two different conditions, we can write Charles' law as follows:

$$\frac{V_1}{T_1} = \frac{V_1}{T_2}$$

Absolute Zero

In this experiment, the volume of a gas will be measured at different temperatures. The gas sample will be the amount of air contained in a 125-mL Erlenmeyer flask. A high temperature of about 100°C will be obtained by heating the flask and its air sample in boiling water. The lower temperatures for the gas sample will be obtained by placing the flask with the gas sample in pans of cool water, including one with ice water.

Absolute zero is the theoretical value for the coldest temperature that matter can attain. Using the data for volume of a gas at different temperatures, a graph will be prepared that shows a relationship between decreasing volume and decreasing temperature. The value of absolute zero is predicted by extending the graph line (extrapolating) to the axis where the volume of the gas would decrease to 0 mL.

Lab Information

Time: 2 hr
Comments: Be careful when you work around boiling water.
               Tear out the report sheets and place them beside the matching procedures.
Related Topics: Pressure, volume, Boyle's law, Charles' law, combined gas law, Kelvin temperature
               ideal gas law

Experimental Procedures

Goggles are required!

A. Boyle's Law

A.1 In an experiment, the volume of a specific amount of gas is measured at different volumes while the temperature is kept constant. You will work with the results given in the report sheet table. Determine the $P \times V$ product by multiplying the pressure and the volume in each sample. Round off the product to give the correct number of significant figures.

A.2 Review the graphing instructions on page xviii in the preface. Use as much of the graph area as possible. Mark the vertical axis in equal intervals of mm Hg of pressure. Divide the horizontal axis into equal intervals of mL. For the lowest pressure value, use a pressure that is slightly below the lowest value in the data: the highest value should be just above the highest measured value for pressure. For example, the pressure scale might begin at 600 mm Hg and go up to 1000 mm Hg. The volume values also should begin with a value near the smallest volume obtained.

Plot the pressure (mm Hg) of the gas in each reading against the volume (mL). Draw a smooth line through the points obtained from the data. The data points will fall on a slight curve called a hyperbola, not a straight line. Use the graph to discuss the meaning of Boyle's law.

B. Charles' Law

Materials: 125-mL Erlenmeyer flask, 400-mL beaker, one-hole rubber stopper with a short piece of glass tubing inserted and attached to a piece of rubber tubing, water containers, thermometer, pinch clamp, buret clamp, Bunsen burner (or a hot plate), wire gauze, iron ring and stand, a graduated cylinder, boiling chips, ice

Dry any moisture on the inside of a 125-mL Erlenmeyer flask. Place a one-hole rubber stopper and tubing in the neck of the flask. (Be sure it has one hole.) Set a 400-mL beaker on a hot plate or on an iron ring covered by wire gauze. Add a few boiling chips to the bottom of the beaker. Attach a buret clamp to the neck of the flask and lower the flask into a 400-mL beaker without touching the flask to the bottom. Fasten the buret clamp to a ring stand. Pour water into the beaker until it comes up to the neck of the flask. Leave space at the top for the water to boil without boiling over. See Figure 13.1 for an illustration of one way set up a boiling water bath.

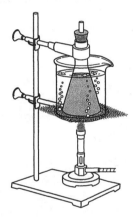

Figure 13.1 Setup for heating an Erlenmeyer flask in a boiling water bath

B.1 Begin heating and bring the water to a boil. Boil gently for 10 minutes to bring the temperature of the air in the flask to that of the boiling water. When you are ready to remove the flask, measure the temperature of the boiling water. Convert from degrees Celsius to the corresponding Kelvin temperature.

Turn off the burner and immediately place a pinch clamp on the rubber tubing. Undo the buret clamp from the ring stand. In the lab, there should be 3 or 4 large containers with water at different temperatures. Using the buret clamp like a handle, *carefully* lift the flask out of the hot water and carry it over to one of the cool water containers. *Keeping the stopper end of the flask pointed downward,* immerse the flask in the cool water. See Figure 13.2. Keeping the flask *inverted,* remove the pinch clamp. Water will enter the flask as the air sample cools and decreases in volume.

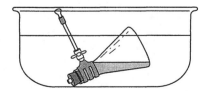

Figure 13.2 Place the heated flask in a pan of cool water

B.2 After the flask has been immersed in the cool water for at least 10 minutes, measure the temperature of the water in your cool water bath. We assume that the temperature of the cooled air sample in the flask is the same as the cool water outside the flask. Record. Convert the Celsius temperature to kelvins.

131

B.3 *Keeping the flask inverted (upside down),* raise or lower the flask until the water level *inside* the flask is equal to the water level in the container. See Figure 13.3. While holding the flask at this level, reattach the pinch clamp to the rubber tubing. Remove the flask from the water, and set it upright on your desk. Remove the clamp and use a graduated cylinder to measure the volume of water that entered the flask as the air sample cooled. Record the volume in mL.

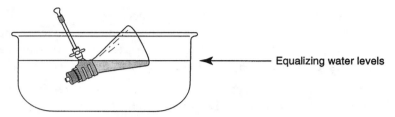

← ———— Equalizing water levels

Figure 13.3 Equalize the water levels inside and outside the inverted flask.

B.4 Measure the total volume (mL) of the flask by filling it to the top with water, but leave room for the volume occupied by the rubber stopper assembly. Record.

B.5 List the temperatures (K) of the boiling water bath and the total gas volume (mL) obtained by other students in the lab.

B.6 Calculate an *average* for the boiling water bath temperature (K).

B.7 Calculate an *average* for the total gas volume (mL) of the flask.

B.8 From the other students in the class, obtain the Kelvin temperatures for 4 other cool water baths and the volume of water that entered each of the other flasks. Use the *average* temperature for the temperature of the boiling water bath. Calculate the volume of air in the flask for each temperature by subtracting the amount of cool water that entered the flask from the *average* total gas volume.

Volume of cooled air = total flask volume (average) – volume of water in flask

B.9 For each gas sample, calculate the V/T_K value for each. These should be constant, according to Charles' law. Any value that is not similar will not be a good data point to use on the graph.

B.10 Graph the volume–temperature relationship of the gas. Review the graphing instructions on page xviii in the preface. The temperature (K) scale starts at 0 K and goes to 400 K. The volume axis extends from 0 mL to 150 mL. Theoretically, gases would reach a temperature called *absolute zero* at a volume of 0 mL. Draw the best straight line you can through the points you plotted on the graph. Then extend the line so it goes all the way to the temperature axis, which equals a volume of zero (0). Your prediction of absolute zero is the temperature where your extended line crosses the horizontal axis. Record this value. See Figure 13.4.

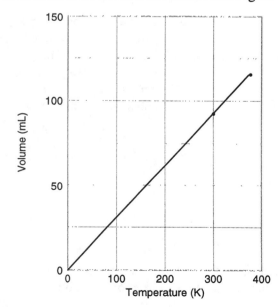

Figure 13.4 Graph of temperature versus the volume of a gas

Report Sheet - Lab 13

Date _____   Name _____

Section _____   Team _____

Instructor _____

Pre-Lab Study Questions

1. What are some occasions that require you to measure the pressure of a gas?

2. Why is an airplane pressurized?

3. Why does a scuba diver need increased gas pressure in the air tank?

4. How does temperature affect the kinetic energy of gas molecules?

A.   Boyle's Law

A.1

Reading	Pressure (P)	Volume (V)	$P \times V$ (Product)
1	630 mm Hg	32.0 mL	
2	690 mm Hg	29.2 mL	
3	726 mm Hg	27.8 mL	
4	790 mm Hg	25.6 mL	
5	843 mm Hg	24.0 mL	
6	914 mm Hg	22.2 mL	

How do the values for the $P \times V$ product compare to each other?

Report Sheet - Lab 13

A.2 Graphing Pressure and Volume: Boyle's Law

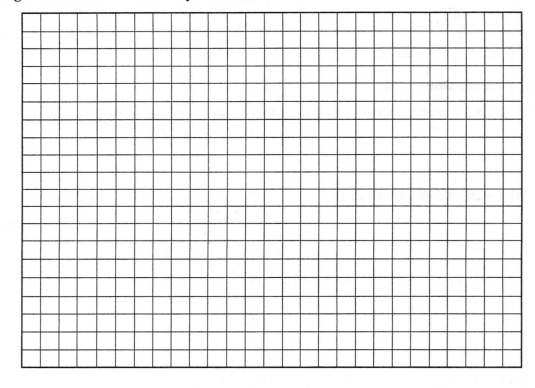

Volume (mL)

Questions and Problems

Q.1 According to your graph, what is the relationship between pressure and volume?

Q.2 On your graph, what is the volume of the gas at a pressure of 760 mm Hg?

Q.3 On your graph, what is the pressure of the gas when the volume is 30.0 mL?

Q.4 Complete the following when T and n remain constant:

Pressure	Volume
Increases	_____
_____	Increases
Decreases	_____

Report Sheet - Lab 13

Q.5   A sample of helium has a volume of 325 mL and a pressure of 655 mm Hg. What will be the pressure if the helium is compressed to 125 mL (*T* constant)? (*Show work.*)

Q.6   A 75.0-mL sample of oxygen has a pressure of 1.50 atm. What will be the new volume if the pressure becomes 4.50 atm (*T* constant)? (*Show work.*)

Q.7   In a weather report, the atmospheric pressure is given as 29.4 inches of mercury. What is the corresponding pressure in mm Hg?

B.   Charles' Law

B.1   Temperature of boiling water bath   _____°C

_____ K

B.2   Temperature of cool water (bath _____)   _____°C

_____ K

B.3   Volume of cool water that entered flask   _____ mL

B.4   Total gas volume of flask   _____ mL

Report Sheet - Lab 13

B.5

Temperature of boiling water baths		Volume (mL) of flasks
°C	K	

B.6 Average boiling water bath temperature (K) _____ K

B.7 Average total gas volume in flask _____ mL

B.8

Water bath	Temperature of water bath		Average total gas volume	Cool water that entered flask	Volume of air remaining in flask
	°C	K			
Boiling water bath				0.0 mL	
Cool bath 1					
Cool bath 2					
Cool bath 3					
Cool bath 4					
Cool bath 5					

B.9

	Boiling water bath	Bath 1	Bath 2	Bath 3	Bath 4	Bath 5
Volume (mL)						
Temperature (K)						
Volume Temperature						

Report Sheet - Lab 13

B.10 Graphing Volume and Temperature Relationship

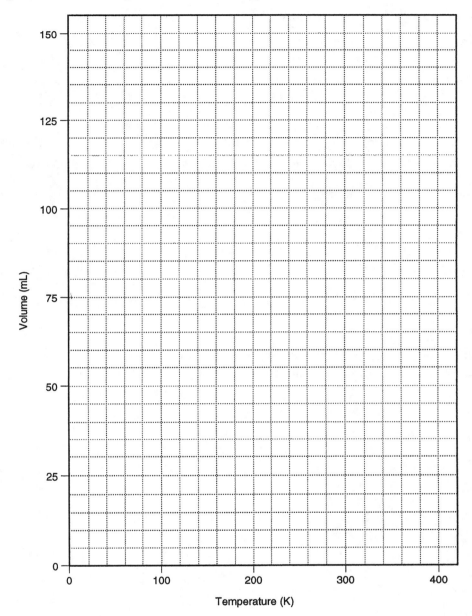

According to your graph, what is the predicted Kelvin temperature of absolute zero?

How does your predicted value for absolute temperature compare with the accepted value of 0 K?

Q.8 Using your graph, how are temperature and volume related?

Q.9 Indicate the change expected when pressure is held constant for a given amount of gas.

Volume	**Temperature**
Increases	_____
_____	Decreases
Decreases	_____
_____	Increases

Q.10 A gas with a volume of 525 mL at a temperature of –25°C is heated to 175°C. What is the new volume of the gas if pressure and number of moles are held constant?

Q.11 A gas has a volume of 2.8 L at a temperature of 27°C. What temperature (°C) is needed to expand the volume to 15 L? (*P* and *n* are constant.)

Q.12 Combined gas law problem: A balloon is filled with 500.0 mL of helium at a temperature of 27°C and 755 mm Hg. As the balloon rises in the atmosphere, the pressure and temperature drop. What volume will it have when it reaches an altitude where the temperature is –33°C and the pressure is 0.65 atm?

Partial Pressures of Oxygen, Nitrogen, and Carbon Dioxide

Goals

- Measure the percentage of O_2 and N_2 in the air.
- Determine the partial pressures of O_2 and N_2 in air.
- Determine the partial pressure of CO_2 in air.
- Determine the partial pressure of CO_2 in expired (exhaled) air.

Discussion

A. Partial Pressures of Oxygen and Nitrogen in Air

The two major gases in the air are oxygen (O_2) and nitrogen (N_2). In this experiment, you will make measurements to use in calculating the percentage of nitrogen and oxygen in the air. Removing the oxygen from a sample of air and measuring the change in volume will determine the amount of oxygen in air. To remove the oxygen, we place some iron filings in a moistened test tube and allow them to react with the oxygen.

$$4Fe \ + \ 3O_2 \ \longrightarrow \ 2Fe_2O_3$$

As the oxygen reacts, the volume it occupied in the air sample is replaced by water. By measuring the volume of the air in the container, and the remaining volume of nitrogen, the volume of oxygen and its percentage in air can be calculated. Using the atmospheric pressure, the partial pressures (mm Hg) of oxygen gas and nitrogen gas in the air can be calculated. *Partial pressures* are the individual pressures exerted by each of the gases that make up the total atmospheric pressure.

B., C. Carbon Dioxide in the Atmosphere and Expired Air

When the body metabolizes nutrients, one of the end products is carbon dioxide, CO_2. The level of carbon dioxide in the blood triggers breathing mechanisms and provides for the correct pH of the blood. Accumulation of carbon dioxide above these levels can result in respiratory and metabolic dysfunction and possible death. The body eliminates most of the carbon dioxide by exhalation of air from the lungs. Plants utilize most of the CO_2 in the atmosphere and return O_2 to the atmosphere.

In experiments B and C, the partial pressures of carbon dioxide in the atmosphere and in expired (exhaled) air will be determined by reacting CO_2 with NaOH.

$$CO_2(g) \ + \ NaOH \ \longrightarrow \ NaHCO_3$$

The other gases in the atmosphere, oxygen and nitrogen, do not react with sodium hydroxide and continue to exert their respective partial pressures.

Lab Information

Time: 2 hr (part A is finished on the next lab day)
Comments: Tear out the report sheets and place them beside the matching procedures.
Related Topics: Partial pressure, Dalton's law, gas mixtures, atmospheric pressure, pressure gradient

Experimental Procedures

GOGGLES REQUIRED!

A. Partial Pressures of Oxygen and Nitrogen in Air

Materials: 250-mL beaker, large test tube, iron filings, graduated cylinder, clamp or test tube holder

A.1 Completely fill a large test tube with water. Measure the volume of the water in the test tube by emptying the water into a graduated cylinder. This is equal to the volume of air in the test tube. Obtain a small scoop of iron filings and sprinkle them in the empty but moist test tube. Shake the iron filings about the test tube. Some should adhere to the sides. Shake out any loose filings in the test tube.

Fill a 250-mL beaker about one-half full of water. Place the test tube containing the iron filings upside down in the water. Attach a clamp or test tube holder to the test tube and let the handle rest on the rim of the beaker. This will stabilize the inverted test tube. Carefully store the beaker and test tube in a place where they will be undisturbed. This part of the experiment will be finished at the beginning of the next laboratory period. See Figure 14.1.

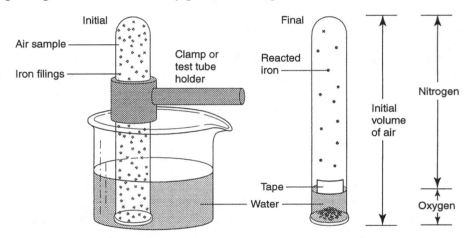

Figure 14.1 Beaker and test tube assembly for determination of the partial pressures of oxygen and nitrogen in the air

A.2 *Next laboratory period* At your next laboratory class, you should find that water has filled part of the test tube. This happened because the iron filings in the test tube reacted with the oxygen in the air. As the oxygen is used up, it is replaced by water. Use a marking pen or tape to mark the water level inside the test tube. Then remove the test tube from the beaker.

Fill the test tube with water *up to the line* you marked. Empty the water into a graduated cylinder and record its volume. This volume is equal to the volume of the nitrogen in the air that *remains* in the test tube after the oxygen reacted.

A.3 Read a barometer and record the atmospheric pressure (mm Hg).

Calculations

A.4 Calculate the volume (mL) of oxygen initially present in the test tube by subtracting the volume (mL) of nitrogen gas from the total volume of the test tube.

$$\text{Volume } (O_2) = \text{Volume of test tube} - \text{Volume of nitrogen } (N_2)$$

A.5 Calculate the percentages of oxygen and nitrogen in air by dividing the O_2 (or N_2) volume by the total volume of the initial air sample.

$$\frac{\text{Volume } O_2 \text{ (or } N_2)}{\text{Volume of air}} \times 100 = \% \ O_2 \text{ (or } N_2) \text{ in air}$$

A.6 Calculate the partial pressures of oxygen (O_2) and nitrogen (N_2). Multiply the atmospheric pressure by the percentage you calculated for each gas.

$$\text{Atmospheric pressure} \times \frac{\% \ O_2 \ (\text{or} \ N_2)}{100} = \text{partial pressure of } O_2 \ (\text{or} \ N_2)$$

B. Carbon Dioxide in the Atmosphere *(This may be a demonstration by the instructor.)*

Materials: 250-mL Erlenmeyer flask to fit two-hole stopper, shell vial, 6 M NaOH, mineral oil, 150-mL beaker, food coloring (optional), meterstick, glass tubing (60–75 cm), two-hole stopper with two short pieces of glass tubing, rubber tubing (1 short, 1 long), funnel, pinch clamp

B.1 Read a barometer and record the atmospheric pressure (mm Hg).

B.2 Carefully lower an empty shell vial into a 250-mL Erlenmeyer flask so that the vial remains upright. See Figure 14.2.

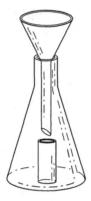

Figure 14.2   Filling a shell vial with NaOH using a funnel in a flask

Set a funnel in the flask directly above the shell vial. Obtain a small amount of 6 M NaOH in a small beaker, and *slowly* pour the NaOH through the funnel into the vial. Stop when the vial is about 3/4 full. Pour a small amount of mineral oil into the vial until the oil forms a *thin* layer of about 1 mm on top. The mineral oil on top of the sodium hydroxide prevents it from reacting with carbon dioxide. Remove the funnel.

Caution: Sodium hydroxide is caustic! Be sure to clean up any spill immediately. Wash any spills on skin for 10 minutes.

Prepare a setup as shown in Figure 14.3. Place the two-holed stopper containing two pieces of glass tubing in the flask. Make sure that the stopper is tight. Attach a short piece of rubber tubing (A) to one piece of glass tubing. Attach a longer piece of rubber tubing to the other glass tubing in the stopper (B). To the open end of the long piece of rubber tubing, attach a long piece (60–70 cm) of glass tubing. Place the open end of the glass tubing in a beaker containing water and a few drops of food coloring (optional). With the rubber tubing open, the flask is full of air at atmospheric pressure. *Close* the system by attaching a pinch clamp to the short piece of rubber tubing (A).

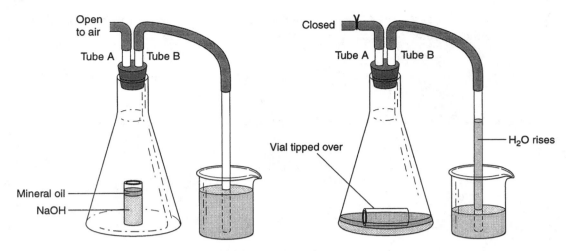

Figure 14.3 Flask, tubing, and vial setup for carbon dioxide determination

Gently tilt the flask and tip over the shell vial. The NaOH solution will spill out into the bottom of the flask. Swirl or shake the flask gently. (Hold the flask around the top with your fingers. Try not to let your hands warm the flask, because an increase in temperature will change the pressure.) *Make sure that the end of tube B stays below the water level in the beaker.* As the carbon dioxide in the sample reacts with NaOH, there is a drop in the pressure of the gases in the sample, and water will rise in the glass tube. *However, because there is only a small amount of CO_2 in the air, the water level will not change very much.*

$$CO_2(g) \quad + \quad NaOH \quad \longrightarrow \quad NaHCO_3$$

Continue to swirl the flask until there is no further change in the water level. *Make sure that the end of tube B remains below the water level in the beaker.* Using a meterstick, measure the distance in millimeters (mm) between the water level in the beaker and the water level in the vertical glass tube B. This is equal to the partial pressure of CO_2 in mm H_2O.

Calculations

B.3 Calculate the P_{CO_2} in mm Hg by dividing the P_{CO_2} (mm H_2O) by 13.6. (1.00 mm Hg exerts the same pressure as 13.6 mm H_2O.) This is equal to the partial pressure of CO_2 (mm Hg).

$$1 \text{ mm Hg} \quad = \quad 13.6 \text{ mm H}_2\text{O}$$

$$\text{Height in mm H}_2\text{O} \quad \times \quad \frac{1 \text{ mm Hg}}{13.6 \text{ mm H}_2\text{O}} \quad = \quad P_{CO_2} \text{ in the atmosphere (mm Hg)}$$

B.4 Calculate the percent CO_2 in the atmosphere by dividing the P_{CO_2} (B.3) by the atmospheric pressure.

$$\frac{P_{CO_2} \text{ (mm Hg)}}{P \text{ atmospheric (mm Hg)}} \quad \times \quad 100 \quad = \quad \% \text{ CO}_2 \text{ in the atmosphere}$$

C. Carbon Dioxide in Expired Air

Materials: 250-mL Erlenmeyer flask to fit two-hole stopper, shell vial, 6 M NaOH, mineral oil, 150-mL beaker, food coloring (optional), meterstick, glass tubing (60–75 cm), two-hole stopper with two short pieces of glass tubing, rubber tubing (1 short, 1 long), funnel, pinch clamp, clean straws

C.1   Read a barometer and record the atmospheric pressure (mm Hg).

C.2   Rinse out the flask you used in part B, then set up the apparatus as you did in part B. This time, place a clean straw into the rubber tubing A. Place glass tube B in the beaker of water.

Take a breath of air, hold for a moment, and exhale through the straw. This will cause bubbling in the beaker of water. Cover the straw with your finger while you inhale. Exhale through the straw again. Repeat this process 3 or 4 times. Place a pinch clamp on the short piece of rubber tubing (A). Remove the straw. **Caution: Take your time. Exhaling too rapidly may cause hyperventilation and make you dizzy. Stop and rest.**

Now the flask is filled with expired (exhaled) air. Repeat the procedure for the CO_2 reaction in the flask with NaOH. Gently tilt the flask and tip over the shell vial, allowing the NaOH to spill out into the bottom of the flask. Swirl or shake the flask gently. (Hold the flask around the top with your fingers. Try not to let your hands warm the flask, because an increase in temperature will change the pressure.) *Make sure that the end of tube B remains below the water level in the beaker.* As the carbon dioxide in the sample reacts with NaOH, there is a drop in the pressure of the gases in the sample, and water will rise in the glass tube. Because of the higher partial pressure of CO_2 in exhaled air, the water level in tube B will make a dramatic rise.

$$CO_2(g) \;+\; NaOH \;\longrightarrow\; NaHCO_3$$

Keep swirling the flask a few minutes until there is no further change in the water level. Using a meterstick, measure the distance (mm) *between* the water level in the beaker and the water level in the vertical glass tube. Make sure your measurement is in millimeters (mm), not centimeters (cm).

Calculations

C.3   Calculate the P_{CO_2} in mm Hg by dividing the P_{CO_2} (mm H_2O) by 13.6. (1 mm Hg exerts the same pressure as 13.6 mm H_2O.) This is equal to the partial pressure of CO_2 (mm Hg) for expired air.

$$1 \text{ mm Hg} = 13.6 \text{ mm } H_2O$$

$$\text{Height in mm } H_2O \;\times\; \frac{1 \text{ mm Hg}}{13.6 \text{ mm } H_2O} \;=\; P_{CO_2} \text{ in expired air (mm Hg)}$$

C.4   Calculate the percent CO_2 in expired air by dividing the P_{CO_2} (C.3) by the atmospheric pressure (C.1).

$$\frac{P_{CO_2} \text{ (mm Hg)}}{P \text{ atmospheric (mm Hg)}} \;\times\; 100 \;=\; \% \; CO_2 \text{ in expired air}$$

Report Sheet - Lab 14

Date _____   Name _____

Section _____   Team _____

Instructor _____

Pre-Lab Study Questions

1.   What is meant by the term *partial pressure*?

A.   Partial Pressures of Oxygen and Nitrogen in Air

A.1   Volume of test tube   _____ mL

A.2   Volume of nitrogen   _____ mL

A.3   Atmospheric pressure   _____ mm Hg

Calculations

A.4   Volume of oxygen   _____ mL

A.5   Percent oxygen   _____ % O_2
      (Show calculations.)

      Percent nitrogen   _____ % N_2
      (Show calculations.)

A.6   Partial pressure of oxygen   _____ mm Hg

      Partial pressure of nitrogen   _____ mm Hg

Questions and Problems

Q.1   If you lived in the mountains where the atmospheric pressure is 685 mm Hg, what would the partial pressure of oxygen and of nitrogen be? Use your % values from A.5.

Report Sheet - Lab 14

B. Carbon Dioxide in the Atmosphere

B.1 Atmospheric pressure   _____ mm Hg

B.2 P_{CO_2} (height of water column in tube B)   _____ mm H_2O

Calculations

B.3 P_{CO_2}

    _____mm H_2O $\times$ $\dfrac{1 \text{ mm Hg}}{13.6 \text{ mm } H_2O}$ =   _____ mm Hg

B.4 Percent CO_2 in the atmosphere   _____% CO_2 (atmosphere)
    (Show calculations.)

C. Carbon Dioxide in Expired Air

C.1 Atmospheric pressure   _____ mm Hg

C.2 Height of water column   _____ mm H_2O

Calculations

C.3 P_{CO_2}

    _____mm H_2O $\times$ $\dfrac{1 \text{ mm Hg}}{13.6 \text{ mm } H_2O}$ =   _____ mm Hg

C.4 Percent CO_2 (expired air)   _____ % CO_2 (expired air)
    (Show calculations.)

Questions and Problems

Q.2 Which would you expect to be higher, the percentage of CO_2 in the atmosphere or in expired air? Why?

Q.3 What is the total pressure in mm Hg of a sample of gas that contains the following gases: O_2 (45 mm Hg), N_2 (1.20 atm), and He (825 mm Hg)?

Q.4 A mixture of gases has a total pressure of 1650 mm Hg. The gases in the mixture are helium (215 mm Hg), nitrogen (0.28 atm), and oxygen. What is the partial pressure of the oxygen in atm?

Solutions, Electrolytes, and Concentration

Goals

- Observe the solubility of a solute in polar and nonpolar solvents.
- Determine the effect of particle size, stirring, and temperature on the rate of solution formation.
- Identify an unsaturated and a saturated solution.
- Compare the conductivity of strong electrolytes, weak electrolytes, and nonelectrolytes.
- List the electrolytes and their concentrations (mEq/L) in intravenous solutions.
- Calculate the mass/mass percent and mass/volume percent concentrations for a NaCl solution.
- Calculate the molar concentration of the NaCl solution.

Discussion

A. Polarity of Solutes and Solvents

A solution is a mixture of the particles of two or more substances. The substance that is present in the greater amount is called the *solvent*. The substance that is present in the smaller amount is the *solute*. In many solutions, including body fluids and the oceans, water is the solvent. Water is considered the *universal solvent*. However, the solutes and solvents that make up solutions may be solids, liquids, or gases. Carbonated beverages are solutions of CO_2 gas in water.

A solution forms when the attractive forces between the solute and the solvent are similar. A polar (or ionic) solute such as NaCl is soluble in water, a polar solvent. As the NaCl dissolves, its ions separate into Na^+ and Cl^-. The positive Na^+ ions are attracted to the partially negative oxygen atoms of water. At the same time, the negative Cl^- ions are pulled into the solvent by their attraction to the partially positive hydrogen atoms of water. Once the ions are into the solvent, they stay in solution because they are hydrated, which means that a group of water molecules is attracted to each ion.

Water, which is polar, dissolves polar solutes such as glucose and salt, NaCl. A nonpolar solvent such as acetone is needed to dissolve a nonpolar solute such as nail polish. This requirement of similar electrical attraction between solute and solvent is sometimes stated as "like dissolves like."

B. Electrolytes and Conductivity

The types of substances in aqueous solutions can be identified as strong electrolytes, weak electrolytes, or nonelectrolytes by using a conductivity apparatus. Electrolytes are substances that produce ions in water. Strong electrolytes contain only ions in solution; weak electrolytes produce a few ions, but contain mostly molecules. Nonelectrolytes dissolve as molecules.

When ions are present in an aqueous solution, the light bulb of a conductivity apparatus will glow because the ions complete an electrical circuit. A nonelectrolyte produces only molecular substances, which do not carry current in an aqueous solution. The light bulb in the conductivity apparatus does not glow. Weak electrolytes produce a few ions: the light bulb will glow weakly. See Figure 15.1.

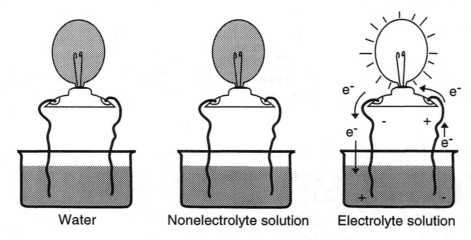

Figure 15.1 Light bulb apparatus for conductivity determination.

C. Electrolytes in Body Fluids

The cells of the body are bathed both inside and outside by fluids that contain specific, differing amounts of electrolytes. The electrolytes play an important role in the maintenance of the activities in the cell. Positive ions are called cations and negative ions are anions. Inside the cell, the major cation is potassium, K^+, and the major anion is bicarbonate, HCO_3^-. The fluid within the cell is called the intracellular fluid. Outside the cell, in the extracellular fluid, the major cation is sodium, Na^+, and the major anion is chloride, Cl^-. Other electrolytes are found in smaller quantities in both intracellular and extracellular fluids.

When there is a loss of fluid from the body or an imbalance of electrolytes, a parenteral solution (one given by means other than oral) may be administered. The type of parenteral solution reflects the needs of the cells in the body as determined by laboratory tests of the body fluids. When giving parenteral solutions such as intravenous solutions, keep in mind the effect of the electrolytes in maintaining and regulating fluid balance, muscle tone, and acid–base balance in the body. See Table 15.1. If there is an imbalance, there may also be a shift of water between plasma and tissues or a loss of water accompanied by a shift or loss of essential electrolytes.

Table 15.1 *Effects of Electrolyte Imbalance*

Electrolyte	Low Levels	High Levels
Na^+, Cl^-	Weakness, headache, diarrhea	Edema
K^+	Apathy, cardiac changes	Cardiac arrest, numbness
Ca^{2+}	Numbness in extremities, tetany	Deep bone pain
HCO_3^-	Acidosis (low pH)	Alkalosis (high pH)
Mg^{2+}	Convulsions, disorientation	Dehydration, coma

D. Concentration of a Sodium Chloride Solution

The concentration of a solution is calculated from the amount of solute present in a certain amount of solution. The concentration may be expressed using different units for amount of solute and solution. A *mass/mass percent* concentration expresses the grams of solute in the grams of solution. The *mass/volume percent* concentration of a solution states the grams of solute present in the milliliters of the solution.

$$\text{mass/mass percent (m/m)} = \frac{\text{grams of solute}}{\text{grams of solution}} \times 100$$

$$\text{mass/volume percent (m/v)} = \frac{\text{grams of solute}}{\text{milliliters of solution}} \times 100$$

A *molar* (M) concentration gives the moles of solute in a liter of solution.

$$\text{molarity (M)} = \frac{\text{moles of solute}}{1 \text{ liter of solution}}$$

In this experiment, you will measure a 10.0-mL volume of a sodium chloride solution. The mass of the solution will be determined by weighing the solution in a preweighed evaporating dish. After the sample is evaporated to dryness, it is weighed again. From this data, the mass of the salt (solute) is obtained.

Using the mass of the solute and the mass of the solution, the mass/mass (m/m) percent can be calculated. From the mass of the solute and the volume (mL) of the solution, the mass/volume (m/v) percent can be calculated. To calculate the molarity of the solution, convert the mass of the solute to moles, and the volume to liters (L).

Lab Information

Time: 3 hr

Comments: Some solvents in part A of the experiments are flammable. Do not light any burners. Tear out the report sheets and place them beside the matching procedures. Your instructor will use the conductivity apparatus in a demonstration for part B.

Related Topics: Solute, solvent, formation of solutions, polar and nonpolar solutes, electrolytes, nonelectrolytes, concentrations of solutions

Experimental Procedures

LABORATORY GOGGLES MUST BE WORN!

A. Polarity of Solutes and Solvents

This may be a demonstration by your instructor.

Materials: Test tubes (8), test tube rack, spatulas, stirring rods, $KMnO_4(s)$, $I_2(s)$, sucrose(s), vegetable oil, cyclohexane

A.1 **Solubility of solutes in a polar solvent** Set up four test tubes in a test tube rack. To each test tube, add a few crystals (or a few drops) of a solute: $KMnO_4$, I_2, sucrose, or vegetable oil. To each, add 3 mL of water and stir the mixture with a glass stirring rod. Describe each solute as soluble or not soluble in the water, a polar solvent. Save for comparison to the test tubes in A.2.

A.2 **Solubility of solutes in a nonpolar solvent** To a different set of four test tubes, add a few crystals (or a few drops) of a solute: $KMnO_4$, I_2, sucrose, or vegetable oil. Place 3 mL of cyclohexane, a nonpolar solvent, in each. *Caution: Cyclohexane is __flammable__—do not proceed if any laboratory burners are in use.*

Indicate whether each solute is soluble or not soluble in cyclohexane, a nonpolar solvent. Compare the solubility of the solutes in cyclohexane (A.2) with their solubility in water (A.1). Discard the solutions for A.1 and A.2 in the waste containers provided in the lab, *NOT* in the sink. *Iodine (I_2) can burn the skin. Handle cautiously!*

A.3 Determine the polarity of each solute from its solubility in each type of solvent. If a solute dissolves in a polar solvent like water, it is a polar solute. If a solute dissolves in a nonpolar solvent like cyclohexane, the solute is nonpolar.

B. Electrolytes and Conductivity *(This will be an instructor demonstration.)*

Materials: Conductivity apparatus, 100- or 150-mL beakers, NaCl(s), 0.1 M NaOH, 0.1 M HCl, 0.1 M $HC_2H_3O_2$ (acetic acid), 0.1 M NH_4OH (ammonium hydroxide), 0.1 M NaCl, 0.1 M sucrose, 0.1 M glucose, ethanol

B.1 Pour about 15–20 mL of the solutions into small beakers. Carefully lower the electrodes into the solution and observe the light bulb. If it glows, the solution is an electrolyte; if not, it is a nonelectrolyte. If the light is very bright, it is a strong electrolyte. A conductivity meter may be substituted for the light bulb apparatus. Select another substance and record the intensity of the light bulb for it. Repeat the test for each sample. Record the intensity of the light bulb in the laboratory record.

> *Caution: Bare electrodes are a hazard! Skin will conduct an electric current and will cause a shock. Do not touch the electrodes when the light bulb apparatus is plugged in.*

B.2 On the laboratory record, indicate the kinds of particles and determine the type of electrolyte that was in each solution.

C. Electrolytes in Body Fluids

Materials: Bags of IV solutions such as saline solution, Ringer's, etc.

C.1 In the laboratory, you should find a display of bags containing intravenous (IV) solutions. Record the type of solutions in three of these.

C.2 Record the electrolytes present in each solution and their concen-trations (mEq/L). These are listed on the label by their symbols. For example, sodium 47 and chloride 47 means that there are 47 mEq of sodium ion and 47 mEq of chloride ion per liter of the solution.

sodium 47 chloride 47 = Na^+, 47 mEq/L,   and Cl^-, 47 mEq/L

C.3 Calculate the total number of mEq of cations and record; calculate the total number of mEq of anions and record.

C.4 Calculate the overall sum of the positive and negative charges and record.

D. Concentration of a Sodium Chloride Solution

Materials: Hot plate (or Bunsen burner, iron ring, and wire screen), evaporating dish, NaCl solution, 400-mL beaker (to fit evaporating dish), 10-mL graduated cylinder (or 10-mL pipet)

D.1 Weigh a dry evaporating dish. Record the mass. *Do not round off.*

D.2 Using a 10.0-mL graduated cylinder, or a 10.0-mL pipet, measure out a 10.0-mL sample of the NaCl solution. See Figure 15.2. Record this volume.

Using a pipet: Place the pipet bulb on the upper end of the pipet. Squeeze the bulb about halfway. Place the tapered end of the pipet in the liquid. As the bulb inflates, liquid will move into the pipet. The level of liquid should rise above the volume mark, but not into the bulb. Remove the bulb and quickly cover the pipet with your *index* finger. Adjust the pressure of your finger to slowly drain the liquid until the level is at the volume mark. You may need to practice. With your finger still on the pipet, lift the pipet out of the liquid and move it to the evaporating dish. Lift your finger off the pipet to let the liquid flow out. Some liquid should remain in the tip.

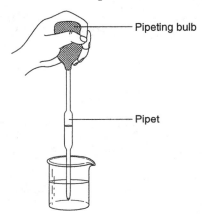

Pipeting bulb

Pipet

Figure 15.2 Using a pipet

D.3 Weigh the evaporating dish and the NaCl solution. Record.

D.4 Fill a 400-mL beaker about half full of water. Set on a hot plate, or heat with a Bunsen burner using an iron ring with a wire screen. Place the evaporating dish on top of the beaker. Heat the water in the beaker to boiling. See Figure 15.3. You may need to add more water to the bath as you proceed.

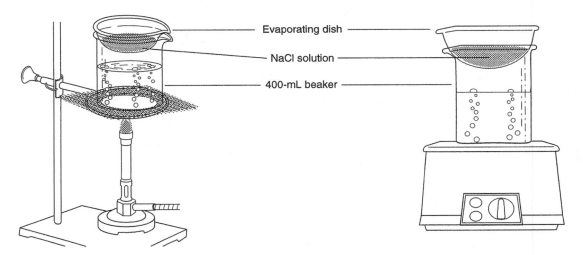

Evaporating dish

NaCl solution

400-mL beaker

Figure 15.3 Using a Bunsen burner or hot plate to evaporate a salt solution

When the NaCl appears to be dry or begins to pop, turn off the burner. After the evaporating dish has cooled, dry the bottom and place it directly on the hot plate or in the iron ring. Heat gently with a low flame to dry the salt completely. Allow the evaporating dish and dried NaCl sample to cool. Weigh the evaporating dish and the dry NaCl. Record. *Do not round off.*

Calculations

D.5 Calculate the mass of the solution.

D.6 Calculate the mass of the NaCl after drying. Subtract the mass of the evaporating dish from the total mass of the evaporating dish and the dried salt.

D.7 Calculate the mass/mass percent concentration.

$$\text{Mass/mass percent} = \frac{\text{Mass of dry NaCl}}{\text{mass (g) of solution}} \times 100$$

D.8 Calculate the mass/volume percent concentration.

$$\text{Mass/volume percent} = \frac{\text{Mass of dry NaCl}}{\text{volume (mL) of solution}} \times 100$$

D.9 Calculate the moles of NaCl. The molar mass of NaCl is 58.5 g/mole.

$$\text{g of dried NaCl} \times \frac{1 \text{ mole NaCl}}{58.5 \text{ g NaCl}} = \text{moles NaCl}$$

D.10 Convert the volume in mL of the solution to the corresponding volume in liters.

$$\text{mL of NaCl solution} \times \frac{1 \text{ L}}{1000 \text{ mL}} = \text{L of NaCl solution}$$

D.11 Calculate the molarity of the NaCl solution.

$$\text{Molarity (M)} = \frac{\text{moles NaCl}}{\text{L of solution}}$$

Report Sheet - Lab 15

Questions and Problems

Q.5   A 15.0-mL sample of NaCl solution has a mass of 15.78 g. After the NaCl solution is evaporated to dryness, the dry salt residue has a mass of 3.26 g. Calculate the following concentrations for the NaCl solution.

a. % (m/m)

b. % (m/v)

c. molarity (M)

Q.6   How many grams of KI are in 25.0 mL of a 3.0 % (m/v) KI solution?

Q.7   How many milliliters of a 2.5 M $MgCl_2$ solution contain 17.5 g $MgCl_2$?

Soluble and Insoluble Salts

Goals

- Predict the formation of an insoluble salt.
- Observe the effect of temperature on solubility.
- Measure the solubility of KNO_3 at various temperatures, and graph a solubility curve.
- Test a variety of water samples for water hardness.
- Use water treatment techniques to purify water.

Discussion

A. Soluble and Insoluble Salts

Although many ionic compounds (salts) are soluble in water, some do not dissolve. They are known as insoluble salts. In medicine, the insoluble salt $BaSO_4$ is used as an opaque substance to help outline the gastrointestinal tract in x-ray images. Solubility rules are shown in Table 16.1.

Table 16.1 *Solubility Rules for Ionic Compounds*

Soluble in Water	Insoluble in Water
Any salt with Li^+, Na^+, K^+, NH_4^+, NO_3^-	
Most chlorides, Cl^-	$AgCl$, $PbCl_2$, and Hg_2Cl_2
Most sulfates, SO_4^{2-}	$BaSO_4$, $PbSO_4$, and $CaSO_4$
	Salts with OH^-, CO_3^{2-}, S^{2-}, PO_4^{3-}

When solutions of two ionic compounds are mixed, the formation of a solid indicates an insoluble salt. The positive ion of one substance and the negative ion of the other formed the insoluble salt. For example, mixing solutions of the soluble salts NaCl and $AgNO_3$ will produce a white solid, which is the insoluble salt AgCl(*s*).

$$AgNO_3(aq) + NaCl(aq) \longrightarrow AgCl(s) + NaCl(aq)$$
$$\underbrace{\hspace{4cm}}_{Soluble\ salts} \qquad \underbrace{\hspace{2cm}}_{Insoluble\ salt}$$

B. Solubility of KNO_3

When a solution holds the maximum amount of solute at a certain temperature, it is *saturated*. When more solute is added, the excess appears as a solid in the container. The maximum amount of solute that dissolves is called the *solubility* of that solute in that solvent. Solubility is usually stated as the number of grams of solute that dissolve in 100 mL (or 100 g) of water. The solubility depends upon several factors, including the nature of the solute and solvent, the temperature, and the pressure (for a gas). Most solids are more soluble in water at higher temperatures. Generally, the dissolving of a solid solute is endothermic, which means that solubility increases with an increase in temperature.

C. Testing the Hardness of Water

Hard water contains the ions Ca^{2+}, Mg^{2+}, and Fe^{3+}. When hard water reacts with soap, the ions in the hard water and some of the soap molecules form insoluble salts called soap scum. The soap molecules tied up in the scum are not free to perform their cleaning function. Initially soap is used to remove the ions, and then more soap is added to produce sudsing and cleaning. The reaction of a soap solution with the ions in hard water can be used to compare hardness of water samples.

D. Purification of Water

In water treatment plants, chemicals added to water will cause the formation of insoluble substances that sink to the bottom of the tank. The water on top is purified and can be drawn off for use.

Lab Information

Time: $2^1/_2$ hr
Comments: Tear out the Lab report sheets and place them beside the matching procedures.
Related Topics: Solubility, insoluble salts, saturated solution

Experimental Procedures   Wear your safety goggles!

A. Soluble and Insoluble Salts

Materials: Spot plate or transparency sheets
0.1 M solutions of $NaCl$, Na_2SO_4, $Ba(NO_3)_2$, $AgNO_3$, Na_3PO_4, $CaCl_2$, $NaOH$, Na_2CO_3
Droppers, or dropper bottles of the 0.1 M solutions

A.1 Write the ions that are in the two solutions used to form each of the mixtures listed.

A.2 Using the solubility rules, draw a circle around the ions that would form an insoluble salt.

A.3 Obtain a spot plate or a plastic sheet. Make mixtures by placing 2–3 drops of each solution in the same well or on the same spot on the plastic sheet. Look for the formation of a solid (insoluble salt). Record your observations.

A.4 Write the formula of the insoluble salts where solids were formed.

B. Solubility of KNO_3

Materials: Weighing paper or small container, spatula, stirring rod, large test tube, 400-mL beaker, buret clamp, hot plate or Bunsen burner, thermometer, 10-mL graduated cylinder, $KNO_3(s)$

To reduce the amount of KNO_3 used, each group of students will be assigned an amount of KNO_3 to weigh out. The results will be shared with the class.

B.1 Obtain a piece of weighing paper or a small container and weigh it carefully. (Or you may tare the container.)

B.2 Each group of students will be assigned an amount of KNO_3 from 2 to 7 grams. Weigh out an amount of KNO_3 that is close to your assigned amount. For example, if you are assigned an amount of 3 grams, measure out a mass such as 3.10 g or 3.25 g or 2.85 g. It is not necessary to add or remove KNO_3 to obtain exactly 3.00 g. Weigh carefully. Calculate the mass of KNO_3.

> ***Taring a container on an electronic balance:*** The mass of a container on an electronic balance can be set to 0 by pressing the tare bar. As a substance is added to the container, the mass shown on the readout is for the substance only. (When a container is *tared*, it is not necessary to subtract the mass of the beaker.)

B.3   *The temperature at which the KNO$_3$ is soluble is determined by heating and cooling the KNO$_3$ solution.* Place 5.0 mL of water in a large test tube. Add your weighed amount of KNO$_3$. Clamp the test tube to a ring stand and place the test tube in a beaker of water. Use a hot plate or Bunsen burner to heat the water. See Figure 16.1. Stir the mixture and continue heating until all the KNO$_3$ dissolves.

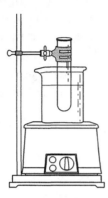

Figure 16.1   Heating the KNO$_3$ solution in a water bath

As soon as all the KNO$_3$ dissolves, turn off the burner or hot plate. Loosen the clamp and remove the test tube from the hot water. As the test tube and contents cool, stir gently with a thermometer. Look closely for the first appearance of crystals. As soon as you see some solid crystals, read the temperature of the solution. Record. This is the temperature at which the solution becomes saturated. The amount of KNO$_3$ in that solution is the solubility of KNO$_3$ at that temperature.

Place the test tube back into the hot water bath and begin heating again. Repeat the warming and cooling of the solution until you have obtained three or more temperature readings that agree. Set the test tube aside. In 15–20 minutes, observe the appearance of the crystals in the test tube.

To discard, add water and heat until the KNO$_3$ dissolves. Pour the solution in proper waste containers provided in the laboratory, *NOT* in the sink. (Solid KNO$_3$ can be recovered from the solution by evaporation to dryness.)

Calculations

B.4   Solubility is expressed as the number of grams of solute in 100 mL of water. Because you used a sample of 5.0 mL of water, the mass of the solute you measured out and the 5.0 mL of water are both multiplied by 20.

$$\frac{g\ KNO_3}{5.0\ mL\ water} \times \frac{20}{20} = \frac{g\ KNO_3}{100\ mL\ water} = \text{Solubility (g KNO}_3 \text{ per 100 mL water)}$$

Collect the solubility results of other KNO$_3$ solutions and their solubility temperatures from the other groups of students in the lab.

B.5   Prepare a graph of the solubility curve for KNO$_3$. Plot the solubility (g KNO$_3$/100 mL water) on the vertical axis and the temperature (0–100°C) on the horizontal axis.

C. Testing the Hardness of Water

Materials: 250-mL beaker, 50-mL (or 25-mL) graduated cylinder, two 250-mL flasks with stoppers, soap solution (dropper bottles), water samples

C.1 Place 50 mL of the water sample you are going to test (begin with distilled water) in a 250-mL flask. Add 1 drop of the soap solution. Stopper the flask and shake for 10 sec. With distilled water, you should see a thick layer of suds. If you don't, add another drop of soap solution and shake for 10 sec again. The suds that form in the distilled water sample will serve as your reference sample. Save for comparison. Shake again if necessary.

Add 1 drop of soap solution to another water sample. Shake. If no suds form, add drops of soap solution until the sample forms an amount of suds similar to that in the distilled water sample. Stop if no suds are formed after you have added 20 drops of soap solution. Test an assortment of water samples available in the lab or from your home, pool, or well. Record the number of drops required to soften each water sample.

D. Purification of Water

Materials: Test tubes, rack, muddy water, 1% $Al_2(SO_4)_3$, 1% Na_2SO_4, 1% NaCl

At water treatment plants, adding chemicals that cause the formation of large particles that sink to the bottom can purify hard water or wastewater. Such a process occurs in the settling tanks at a water filtration facility.

Set up four test tubes in a test tube rack. To each, add 10 mL of muddy water. Add the following chemicals, one to each test tube of the muddy water; label each.

(1) 5 mL water (control)   (3) 5 mL 1% $Al_2(SO_4)_3$

(2) 5 mL 1% NaCl   (4) 5 mL 1% Na_2SO_4

Stir each test tube thoroughly. Then allow the test tubes to stand undisturbed. Look for a separation of a precipitate and a clarification of the upper portion of water. Record your observations after 15, 30, and 45 minutes.

NS   no settling, still muddy   MS mostly settled, slightly cloudy

BS   beginning to settle, still murky   SC settled, clear

SS   some settling, cloudy

Report Sheet - Lab 16

Date _____   Name _____

Section _____   Team _____

Instructor _____

Pre-Lab Study Questions

1. What is an insoluble salt?

A.   Soluble and Insoluble Salts

Compounds in Mixture	A.1, A.2 Ions	A.3 Observations	A.4 Formula of Insoluble Salt (if any)
$AgNO_3$ + NaCl	Ag^+ NO_3^- Na^+ Cl^-		
$AgNO_3$ + Na_2SO_4			
$AgNO_3$ + Na_3PO_4			
$Ba(NO_3)_2$ + Na_2SO_4			
$CaCl_2$ + NaOH			
$CaCl_2$ + Na_2SO_4			
$CaCl_2$ + Na_2CO_3			

Questions and Problems

Q.1   Why do some mixtures of ionic compounds form a solid precipitate?

B.   Solubility of KNO₃

B.1 Mass of Container	B.2 Mass of Container + KNO₃	Mass of KNO₃	B.3 Temperature (Crystals Appear)	B.4 Solubility (g KNO₃/100 mL H₂O)

B.5   Graphing the solubility of KNO₃ vs. temperature (°C)

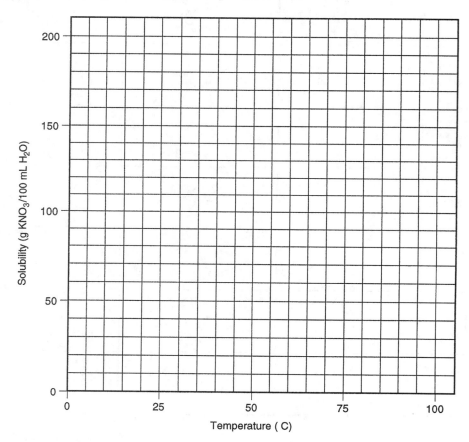

According to your graph, what is the effect of increasing temperature on the solubility of KNO₃?

Report Sheet - Lab 16

Questions and Problems

Q.2   The solubility of sucrose (common table sugar) at 70°C is 320 g/100 g H_2O.
    a. How much sucrose can dissolve in 200 g of water at 70°C?

    b. Will 400 g of sucrose dissolve in a teapot that contains 200 g of water at 70°C? Explain.

C.   Testing the Hardness of Water

C.1

Type of Water	Drops of Soap	Type of Water	Drops of Soap
distilled water	_____	mineral water	_____
tap water	_____	seawater	_____
_____	_____	_____	_____
_____	_____	_____	_____
_____	_____	_____	_____
_____	_____	_____	_____

Questions and Problems

Q.3   Which water sample was the hardest? Why?

Report Sheet - Lab 16

D. Purification of Water

Agent Added	15 Minutes	30 Minutes	45 Minutes
H_2O			
NaCI			
$Al_2(sO_4)_3$			
Na_2SO_4			

Questions and Problems

Q.4 Which chemical produced the most rapid settling? Why?

17

Testing for Cations and Anions

Goals

- Determine the presence of a cation or anion by a chemical reaction.
- Determine the presence of some cations and anions in an unknown salt.

Discussion

Solutions such as milk, coffee, tea, and orange juice contain an assortment of ions. In chemical reactions, these ions give a distinctive flame test, undergo color changes, or form a gas or an insoluble solid. Your observations of the results of a test are the key to identifying those same ions when you test unknown solutions. After you look at a test that produces some change when a particular ion reacts, you will look for the same change in an unknown sample. If the test result is the same, you assume that the ion is present in the unknown. If the test result does not occur then the ion is not in the unknown: the test is negative.

A. Tests for Positive Ions (Cations)

In this experiment, you will observe the tests for the following cations:

Cations	Solution
sodium (Na^+)	0.1 M NaCl
potassium (K^+)	0.1 M KCl
calcium (Ca^{2+})	0.1 M $CaCl_2$
iron (Fe^{3+})	0.1 M $FeCl_3$
ammonium (NH_4^+)	0.1 M NH_4Cl

The presence of Na^+, K^+, and Ca^{2+} is determined by the distinctive color the ions give in flame tests. The calcium ion also reacts with ammonium oxalate, $(NH_4)_2C_2O_4$, to give a white precipitate. When NH_4^+ is converted to ammonia (NH_3), a distinctive odor is emitted, and the fumes turn red litmus paper to blue. Iron (Fe^{3+}) is detected by the distinctive red color it gives with potassium thiocyanate, KSCN.

B. Tests for Negative Ions (Anions)

In this experiment, you will observe the tests for the following anions:

Anions	Solution
chloride (Cl^-)	0.1 M NaCl
phosphate (PO_4^{3-})	0.1 M Na_3PO_4
sulfate (SO_4^{2-})	0.1 M Na_2SO_4
carbonate (CO_3^{2-})	0.1 M Na_2CO_3

When $AgNO_3$ is added to the solutions to be tested for anions, several insoluble salts form. However, when nitric acid, HNO_3, is added to these precipitates, all the solids except AgCl will dissolve; the insoluble salt AgCl will remain in the test tube. Phosphate ion (PO_4^{3-}) reacts with ammonium molybdate and forms a yellow solid. When barium chloride, $BaCl_2$, is added to a solution containing SO_4^{2-}, a precipitate of $BaSO_4$ forms. Barium may form insoluble salts with some other anions, but the addition of HNO_3 dissolves all barium salts except $BaSO_4$. The insoluble $BaSO_4$ remains in the test tube after HNO_3 is added. Carbonate anion (CO_3^{2-}) is identified by adding HCl, which produces bubbles of CO_2 gas.

C. Writing the Formula of Your Unknown Salt

As you proceed, you will test known solutions that contain a particular ion. Each student will receive an unknown sample containing a cation and an anion. After you observe reactions of knowns, you will carry out the same tests with your unknown. Therefore, you should expect to see a test that matches the reactions of one of the cations, and another for one of the anions. By identifying the ions in your unknown, you will be able to write the name and formula of your unknown salt.

D. Testing Consumer Products for Some Cations and Anions

Consumer products contain many of the same ions that you will test for in the experiment. Once you have gone through the procedures and identified positive test results for cations and anions, the same procedures can be applied to consumer products to identify some of the ions present. After you carry out the tests for cations and anions, you may prepare a solution of a consumer product, and carry out the same tests.

Lab Information

Time: 2½ hr

Comments: Use several small beakers to hold the reagents. Be sure to label each.
                HCl and HNO₃ are strong acids, and NaOH is a strong base. Handle with care!
                If they are spilled on the skin, rinse thoroughly with water for 10 minutes.
                Dispose of test results properly.
                Tear out the report sheets and place them beside the matching procedures.

Related Topics: Ions, chemical change, solubility rules

Experimental Procedures

WEAR YOUR GOGGLES!

A. Tests for Positive Ions (Cations)

Materials: Spot plate, Bunsen burner, test tubes and test tube rack, flame-test wire, 4 small beakers, 3 M HCl, 6 M HNO₃, 6 M NaOH, dropper bottles of 0.1 M NaCl, 0.1 M KCl, 0.1 M CaCl₂, 0.1 M (NH₄)₂C₂O₄ (ammonium oxalate), 0.1 M NH₄Cl, 0.1 M FeCl₃, 0.1 M KSCN (potassium thiocyanate), red litmus paper, warm water bath, stirring rod

In many parts of this experiment, you will be using 2 mL of different solutions. Place 2 mL of water in the same size test tube that you will be using. As you obtain solutions for the experiment, take a volume that matches the height of the 2 mL of water. Be sure to label each test tube. Many test tubes have a frosted section to write on. If not, use a marking pencil or a label. Before you begin this group of experiments, place about 15 mL of each of the following reagents in small beakers and label them: 3 M HCl, 6 M HNO₃, and 6 M NaOH. Another beaker of distilled water is convenient for rinsing out the droppers.

Preparation of an unknown

Take a small beaker to your instructor for a sample of the unknown salt solution assigned to you. Record the sample number. If the unknown is a solid, dissolve 1 g of the salt in 50 mL of distilled water. Use small portions of this *unknown* solution in each of the tests. You will need to identify the presence of a cation and an anion in the unknown by comparing the test results of the unknown with the test results given by each of the known solutions.

A.1 **Flame tests for Na^+, K^+, and Ca^{2+}**

Obtain a spot plate and place 5–8 drops of 0.1 M solutions of NaCl, KCl, $CaCl_2$, and your unknown solution into separate wells. Dip the test wire in 3 M HCl and heat the wire until the flame is light blue. Place the wire loop in the NaCl solution and then into the flame. Record the color produced by the Na^+ ion. Clean the wire, and repeat the test with the KCl solution. The color (pink-lavender) of K^+ does not last long, so look for it immediately. Record the color for K^+. Clean the wire again, and repeat the flame test with Ca^{2+}. Record the color of the Ca^{2+} flame.

Testing the unknown Clean the flame-test wire and dip it in the unknown solution. Record the results when the wire is placed in a flame. If there is a color that matches the color of an ion you tested in the flame tests (A.1), you can conclude that you have one of the ions Na^+, K^+, or Ca^{2+} in your unknown. Record. If you think Ca^{2+} is present, you may wish to confirm it with test A.2. If the flame test does not produce any color, then Na^+, K^+, and Ca^{2+} are not present in your unknown.

A.2 **Test for calcium ion, Ca^{2+}**

Place 2 mL of 0.1 M $CaCl_2$ in a test tube and 2 mL of your unknown solution in another test tube. Add 15 drops of ammonium oxalate solution, 0.1 M $(NH_4)_2C_2O_4$, to each. Look for a cloudy, white solid (precipitate). If the solution remains clear, place the test tube in a warm water bath for 5 minutes, then look for a precipitate. The net equation for the reaction is

$$Ca^{2+} + C_2O_4^{2-} \longrightarrow CaC_2O_4(s)$$
A white precipitate indicates the presence of Ca^{2+}.

A.3 **Test for ammonium ion, NH_4^+**

Place 2 mL of 0.1 M NH_4Cl in a test tube and 2 mL of your unknown in another test tube. Add 15 drops of 6 M NaOH to each. *Carefully* fan the vapors from the test tube toward you. You may notice the odor of ammonia. Place a strip of moistened red litmus paper across the top of the test tube and set the test tube in a warm water bath. The $NH_3(g)$ given off will turn the red litmus paper blue.

$$NH_4^+ + OH^- \longrightarrow NH_3(g) + H_2O$$
Ammonia

Repeat the test with your unknown. Record results.

A.4 **Test for ferric ion, Fe^{3+}**

Place 2 mL of 0.1 M $FeCl_3$ in a test tube and 2 mL of your unknown in another test tube. Add 5 drops of 6 M HNO_3 and 2–3 drops of potassium thiocyanate, 0.1 M KSCN. A deep red color indicates that Fe^{3+} is present. A faint pink color is *not* a positive test for iron. Repeat the test with your unknown. Record results.

$$Fe^{3+} + 3SCN^- \longrightarrow Fe(SCN)_3$$
Deep red color

169

B. Tests for Negative Ions (Anions)

Materials: Test tubes, test tube rack, 0.1 M NaCl, 0.1 M $AgNO_3$ (dropper bottle), 3 M HCl, 6 M HNO_3, stirring rod, 0.1 M Na_2SO_4, 0.1 M $BaCl_2$, 0.1 M Na_3PO_4, $(NH_4)_2MoO_4$ (ammonium molybdate reagent), 0.1 M Na_2CO_3, hot water bath

B.1 **Test for chloride ion, Cl^-** Place 2 mL of 0.1 M NaCl solution in a test tube and 2 mL of your unknown in another test tube. To each sample, add 5–10 drops of 0.1 M $AgNO_3$ and 10 drops of 6 M HNO_3. Stir with a glass stirring rod. *Caution: $AgNO_3$ stains the skin.* Any white solid that remains is AgCl(*s*). Any white solids that dissolve with HNO_3 do not contain Cl^-. Record the results of your known and unknown.

$$Ag^+ + Cl^- \longrightarrow AgCl(s)$$
White precipitate remains after HNO_3 is added.

B.2 **Test for sulfate ion, $SO_4{}^{2-}$** Place 2 mL of 0.1 M Na_2SO_4 solution in a test tube and 2 mL of your unknown in another test tube. Add 1 mL (20 drops) of $BaCl_2$ and 5–6 drops of 6 M HNO_3 to each test tube. $BaSO_4$, a white precipitate, does not dissolve in HNO_3. Other anions, $CO_3{}^{2-}$ and $PO_4{}^{3-}$, will also form barium compounds, $Ba_3(PO_4)_2$ and $BaCO_3$, but they will *dissolve* in HNO_3. Record your test results for the known and unknown.

B.3 **Test for phosphate ion, $PO_4{}^{3-}$** Place 2 mL of 0.1 M Na_3PO_4 solution in a test tube and 2 mL of your unknown in another test tube. Add 10 drops of 6 M HNO_3 to each. After the test tubes are warmed in a hot water bath (60°C), add 15 drops of ammonium molybdate solution, $(NH_4)_2MoO_4$. The formation of a yellow precipitate indicates the presence of $PO_4{}^{3-}$. Record the test results of the known and the unknown.

B.4 **Test for carbonate ion, $CO_3{}^{2-}$** Place 2 mL of 0.1 M Na_2CO_3 solution in a test tube and 2 mL of your unknown in another test tube. ***While carefully observing the solution***, add 10 drops of 3 M HCl to each sample. Watch for a strong evolution of bubbles of CO_2 gas as you add the HCl. The gas bubbles are formed quickly, and may be overlooked. If gas bubbles were not observed, add another 15–20 drops of HCl *as you watch the solution*. Record your results for the known and the unknown.

$$Na_2CO_3(aq) + 2HCl(aq) \longrightarrow CO_2(g) + H_2O + 2NaCl(aq)$$
Gas bubbles

C. Writing the Formula of Your Unknown Salt

Your unknown solution was made from a salt composed of a cation and an anion. From your test results, you can identify one of the cations (Na^+, K^+, Ca^{2+}, NH_4^+, or Fe^{3+}) and one of the anions (Cl^-, $SO_4{}^{2-}$, $PO_4{}^{3-}$, or $CO_3{}^{2-}$). For example, if you found that in the cation tests you got the same test result as for Ca^{2+} and in the anion tests you got the same result as for Cl^-, then the ions in your unknown salt would be Ca^{2+} and Cl^-. The formula $CaCl_2$ is written using charge balance.

C.1 Write the symbols and names of the cation and anion that were present in your unknown.

C.2 Use the ionic charges of the cation and anion to write the formula and name of the salt that was your unknown.

D. Testing Consumer Products for Some Cations and Anions

The tests in this experiment may be used to identify ions in samples of the following consumer products. Perhaps you have other ideas for products to test. Ask your instructor. In many of the products, there will be several cations and anions that give positive tests. Describe your results on the lab report.

Product	Procedure
Juices	Obtain 25-30 mL of a light-color fruit juice. If it contains pulp or fiber, filter first. Use 2-3 mL of the final solution for each of the cation and anion tests.
Sodas	Obtain 25-30 mL of a soft drink, or mineral water. For colas, root beers, or others with deep colors, mix the soft drink with a small amount of charcoal in a small beaker. Charcoal will absorb the dyes. Filter and use 2-3 mL of the filtrate for each cation or anion test.
Milk	Obtain 30 mL of nonfat milk. Add 10 mL of 0.1 M acetic acid ($HC_2H_3O_2$). Small particles of protein(curds) will form. Filter. Gently boil the filtrate to reduce the volume to 15-20 mL. Use 2-3 mL of the filtrate for each cation or anion test
Bone meal or plant food granules	In a beaker, mix a scoop of bone meal or plant food with 15 mL distilled water and 15 mL of 6M HNO_3. Heat gently (DO NOT BOIL) until most of the material dissolves. Cool and filter. Use 2-3 mL of the filtrate for each cation or anion test
Window cleaner	Obtain 20-25 mL of a window cleaner. Use 2-3 mL of the filtrate for each cation or anion test

Report Sheet - Lab 17

Date _____   Name _____

Section _____   Team _____

Instructor _____

Pre-Lab Study Questions

1. How can the presence of an ion in a solution be detected?

2. If a reaction produces an insoluble salt, what will you notice happening in the test tube?

A.   Tests for Positive Ions (Cations)

Unknown number _____

Procedure	Cation Tested	Observations	Observations for Unknown
A.1 **Flame tests**	Na^+		
	K^+		
	Ca^{2+}		
A.2 **Oxalate test**	Ca^{2+}		
A.3 **Ammonium test**	NH_4^+		
A.4 **Iron test**	Fe^{3+}		

Identification of the positive ion in the unknown solution
From your test results, what positive ion (cation) is present in your unknown? _____
Explain your choice.

Report Sheet - Lab 17

B. Tests for Negative Ions (Anions)

Procedure	Anion Tested	Observations	Observations for Unknown
B.1 **Chloride test**	Cl^-		
B.2 **Sulfate test**	SO_4^{2-}		
B.3 **Phosphate test**	PO_4^{3-}		
B.4 **Carbonate test**	CO_3^{2-}		

Identification of the negative ion in the unknown solution

From your test results, what negative ion (anion) is present in your unknown? _____
Explain your choice.

C. Writing the Formula of Your Unknown Salt

Unknown sample number _____

C.1 Cation _____   Name _____

Anion _____   Name _____

C.2 Formula of the unknown salt _____

Name of the unknown salt _____

Report Sheet - Lab 17

D. Testing Consumer Products for Some Cations and Anions

Product tested _____

Cation tests	Results	Ion present
Flame tests (Na^+, K^+, Ca^{2+})		
Ca^{2+}		
NH_4^+		
Fe^{3+}		
Anion tests		
Cl^-		
SO_4^{2-}		
PO_4^{3-}		
CO_3^{2-}		

Report Sheet - Lab 17

Questions and Problems

Q.1 How do the tests on known solutions containing cations and anions make it possible for you to identify the cations or anions in an unknown substance?

Q.2 You have a solution that is composed of either NaCl or $CaCl_2$. What tests would you run to identify the compound?

Q.3 If tap water turns a deep red color with a few drops of KSCN, what cation is present?

Q.4 A plant food contains $(NH_4)_3PO_4$. What tests would you run to verify the presence of the NH_4^+ ion and the PO_4^{3-} ion?

Q.5 Write the symbol of the cation or anion that give(s) the following reaction:

_____ a. Forms a precipitate with $AgNO_3$ that does not dissolve in HNO_3

_____ b. Forms a gas with HCl

_____ c. Gives a bright, yellow-orange flame test

_____ d. Forms a precipitate with $BaCl_2$ that does not dissolve in HNO_3

Solutions, Colloids, and Suspensions

Goals

- Perform chemical tests for chloride, glucose, and starch.
- Use dialysis to distinguish between solutions and colloids.
- Separate colloids from suspensions.
- Discuss the effects of hypotonic and hypertonic solutions on red blood cells.

Discussion

A. Identification Tests

A chemical test helps us identify the presence or absence of a substance in a solution. In a *positive test*, the chemical change indicates the substance is present. In a *negative test*, there is no chemical reaction and the substance is absent.

B. Osmosis and Dialysis

Osmosis occurs when water moves through the walls of red blood cells, which are semipermeable membranes. The osmotic pressure of an *isotonic* solution is the same as that of red blood cells. Both 0.9% NaCl (saline) and 5% glucose solutions are considered isotonic to the cells of the body. In isotonic solution, the flow of water into and out of the red blood cells is equal. A *hypotonic* solution has a lower osmotic pressure than an isotonic solution; the osmotic pressure of a *hypertonic* solution is greater. In either case, the flow of water in and out of the cell is no longer equal, and the cell volume is altered.

When red blood cells are placed in a strong salt solution, they form small clumps. This process, called *crenation*, occurs because water diffuses out of the cells into the more concentrated salt solution. If red blood cells are placed in water, they expand and may rupture. This process, called *hemolysis*, occurs because water diffuses into the cells where there is a higher solute concentration. In both cases, osmosis has occurred as water passed through a semipermeable membrane into the more concentrated solution.

In *dialysis*, small particles and water, but not colloids, move across a semipermeable membrane from their high concentrations to low. Many of the membranes in the body are dialyzing membranes. For example, the intestinal tract consists of a semipermeable membrane that allows the solution particles from digestion to pass into the blood and lymph. Larger, incompletely digested food particles that are colloidal size or larger remain within the intestinal wall. Dialyzing membranes are also used in hemodialysis to separate waste particles, particularly urea, out of the blood.

C. Filtration

In the process of *filtration*, gravity separates suspension particles from the solvent. The pores in the filter paper are smaller than the size of the suspension particles, causing the suspension particles to be trapped in the filter paper. The colloidal particles and the solution particles are smaller and can pass through the pores in the filter paper.

In this experiment, a mixture of starch, NaCl, glucose, and charcoal is poured into filter paper. Using identification tests, the substances that remain in the filter paper and the substances that pass through the filter paper can be determined.

Lab Information

Time: $2^{1}/_{2}$ hr

Comments: Label the containers to keep track of different solutions.

                       Tear out the report sheets and place them beside the matching procedures.

Related Topics: Solutions, colloids, suspensions, osmosis, hypertonic solutions, isotonic solutions, hypotonic solutions, dialysis

Experimental Procedures

WEAR YOUR SAFETY GOGGLES!

A. Identification Tests

Materials: Three small beakers (100–150 mL), test tubes (6), test tube rack, stirring rod, 50-mL graduated cylinder, test tube holder, droppers, boiling water bath, 1% starch, 10% NaCl, 0.1 M $AgNO_3$, 10% glucose, Benedict's reagent, iodine reagent. *Dropper bottles with these reagents may be available in the laboratory.*

In this experiment, you will perform the identification tests for Cl^-, glucose, and starch. To determine the presence or absence of Cl^-, glucose, or starch in later experiments, refer to the results of the tests. For each test, observe and record the initial properties of the reagent and the final appearance of the solution after the reagent is added. In a positive test, there will be a change in the original properties of the reagent, such as a color change and/or the formation of a precipitate (opaque solid). If there is no change in the appearance of the reagent, the test is *negative*.

Place small amounts of the reagents for these identification tests in small beakers or vials for use throughout this experiment. Place 3 mL of water in a test tube for volume comparison. Obtain 3–4 mL of 0.1 M $AgNO_3$, 20–25 mL of Benedict's reagent, 4–5 mL of iodine reagent, and 10 mL of distilled water. Label each container. Keep these reagent containers at your desk for the duration of the experiment.

Caution: $AgNO_3$ and iodine reagent stain!

A.1   **Chloride (Cl^-) test**   Place 3 mL of 10% NaCl in a test tube. Place 3 mL of distilled water in another test tube. The test tube with the water will be the control or comparison sample. Test for Cl^- by adding 2 drops of 0.1 M $AgNO_3$ to each. Record your observations. Compare the results for the NaCl solution with the results for the water sample.

A.2   **Starch test**   Place 3 mL of 1% starch solution in a test tube. Place 3 mL of distilled water in another test tube. Add 2–3 drops of iodine reagent to each. Record your observations. Compare the results for the starch sample with the results for the water sample.

A.3   **Glucose test**   Place 3 mL of 10% glucose solution in a test tube. Place 3 mL of distilled water in another test tube. Add 3 mL of Benedict's reagent to the glucose and to the distilled water. Heat both test tubes in a boiling water bath for 5 minutes. Record your observations. Compare the results for the glucose with the results for the water sample. *Note: Each time you carry out the glucose test, the test tubes containing Benedict's reagent must be heated in a boiling water bath.*

B. Osmosis and Dialysis

Materials: Cellophane tube (15–20 cm), test tubes (3), test tube rack, test tube holder, stirring rod, droppers, 50-mL graduated cylinder, funnel, 100-mL beaker, 250-mL beaker, boiling water bath, 10% NaCl, 10% glucose, 1% starch, 0.1 M $AgNO_3$, Benedict's solution, iodine reagent

In a small beaker, combine 10 mL of 10% NaCl, 10 mL of 10% glucose, and 10 mL of 1% starch solution. Tie a knot in one end of a piece of cellophane tubing (dialysis bag). Place a funnel in the open end and pour in about 20 mL of the mixture. Save the rest of the mixture for part C. Tie a firm knot in the open end to close the dialysis bag. Rinse the dialysis bag with distilled water. Place the dialysis bag in a 250-mL beaker and cover with distilled water. See Figure 18.1.

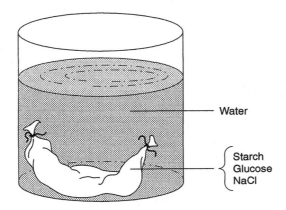

Figure 18.1   Dialysis bag placed in distilled water

Pour off 15 mL of the distilled water surrounding the bag for the first group of tests. Divide the 15 mL into three test tubes (5 mL each). Repeat the tests from part A. A substance is present in the water outside the dialysis bag if a test for that substance gives the same results as in the identification tests in part A. Record the result as positive (+). If the substance is absent (there is no chemical reaction), record that test as negative (–).

Identification Tests:

Test for Cl^-	Add $AgNO_3$ to the dialysis water in the first test tube.
Test for starch	Add iodine reagent to the dialysis water in the second test tube.
Test for glucose	Add Benedict's solution to the dialysis water in the third test tube. Heat.

After 20 minutes and 40 minutes, pour off another 15 mL of water from *outside* the dialysis bag. Separate the water sample into three test tubes and test again for Cl^-, starch, and glucose. Record your test results each time.

After the 40-minute test, break the bag open, and test 15 mL of its contents. From your results, determine which substance(s) dialyzed through the membrane, and which substance(s) did not. *Save the remainder of the bag's contents for use in part C.*

C. Filtration

Materials: Funnel, filter paper, 50-mL graduated cylinder, test tubes (3), test tube rack, test tube holder, stirring rod, droppers, two 150-mL beakers, boiling water bath, powdered charcoal, iodine reagent, 0.1 M $AgNO_3$, Benedict's solution

To the mixture of glucose, NaCl, and starch from part B, add a small amount of powdered charcoal. Fold a piece of filter paper in half and in half again. Open the fold to give a cone-like shape and place the filter paper cone in a funnel. Push it gently against the sides and moisten with water. Place a small beaker under the funnel. Pour the mixture into the filter paper. Collect the liquid (filtrate) that passes through the filter. See Figure 18.2.

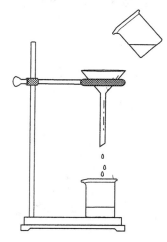

Figure 18.2 Filtering a mixture

C.1 Describe what you see in the filter paper.

C.2 Test the liquid in the beaker (filtrate) for chloride, starch, and glucose using the identification tests from part A. Compare the results of each test to part A. Use the test results to identify the substances that passed through the filter paper.

C.3 From the test results, determine what was trapped in the filter and what passed through. Identify the substance that is a suspension.

Report Sheet - Lab 18

Date _____   Name _____

Section _____   Team _____

Instructor _____

Pre-Lab Study Questions

1. In making pickles, a cucumber is placed in a strong salt solution. Explain what happens.

2. Why is it important that cell membranes are semipermeable membranes?

3. What is the difference between osmosis and dialysis?

4. How does an artificial kidney separate waste products from the blood?

A. Identification Tests

Test	Reagent Added	Results of Positive Test	Results with Water Control
A.1 **Cl⁻**	$AgNO_3$		
A.2 **Starch**	Iodine		
A.3 **Glucose**	Benedict's; heat		

Report Sheet - Lab 18

B. Osmosis and Dialysis

Test Results of Water Outside Dialysis Bag

Time	Cl⁻ Present?	Starch Present?	Glucose Present?
0 minutes			
20 minutes			
40 minutes			
Contents of dialysis bag			

Questions and Problems

Q.1 Which substance(s) were found in the water *outside* the dialysis bag?

Q.2 How did those substance(s) go into the water outside the dialysis bag?

Q.3 What substance(s) were retained inside the dialysis bag? Why were they retained?

C. Filtration

C.1 Appearance of filter paper _____

  Substance present _____

C.2 Test for	Results of Test	Substance Present in Filtrate?
Cl⁻		
Starch		
Glucose		

C.3 Which substance is a suspension? _____

  Which substances are solutions or colloids? _____

Report Sheet - Lab 18

Questions and Problems

Q.4 What is an isotonic solution?

What is a hypotonic solution?

What is a hypertonic solution?

Q.5 State whether each of the following are isotonic, hypotonic, or hypertonic:

a. H_2O _____

b. 0.9% NaCl _____

c. 10% glucose _____

d. 3% NaCl _____

e. 0.2% NaCl _____

Q.6 A red blood cell in a hypertonic solution will shrink in volume as it undergoes *crenation*. In a hypotonic solution, a red blood cell will swell and possibly burst as it undergoes *hemolysis*. Predict the
effect on a red blood cell (crenation, hemolysis, or none) that the following solutions would have:

a. 2% NaCl _____

b. H_2O _____

c. 5% glucose _____

d. 1% glucose _____

e. 10% glucose _____

Q.7 A parenteral solution is a solution that is injected into the tissues or bloodstream, but not given oral-
ly. Why are isotonic solutions used as parenteral solutions?

Acids, Bases, pH, and Buffers

Goals

- Prepare a naturally occurring dye to use as a pH indicator.
- Measure the pH of several substances using cabbage indicator and a pH meter.
- Calculate pH from the $[H^+]$ or the $[OH^-]$ of a solution.
- Calculate the molar concentration and percentage of acetic acid in vinegar.
- Observe the changes in pH as acid or base is added to buffered and unbuffered solutions.
- Calculate pH from the $[H^+]$ or the $[OH^-]$ of a solution.

Discussion

An *acid* is a substance that dissolves in water and donates a hydrogen ion, or proton (H^+), to water. In the laboratory we have been using acids such as hydrochloric acid (HCl) and nitric acid (HNO_3).

$$HCl \;+\; H_2O \;\longrightarrow\; \underset{\textit{hydronium ion}}{H_3O^+} \;+\; Cl^-$$

You use acids and bases every day. There are acids in oranges, lemons, vinegar, and bleach. In this experiment we will use acetic acid ($HC_2H_3O_2$). Acetic acid is the acid in vinegar that gives it a sour taste.

A *base* is a substance that accepts a proton. Some household bases include ammonia, detergents, and oven-cleaning products. Some typical bases used in the laboratory are sodium hydroxide (NaOH) and potassium hydroxide (KOH). Most of the common bases dissolve in water and produce hydroxide ions, OH^-.

$$NaOH \;\longrightarrow\; Na^+ \;+\; OH^-$$

An important weak base found in the laboratory and in some household cleaners is ammonia. In water, it reacts to form ammonium and hydroxide ions:

$$NH_3 \;+\; H_2O \;\longrightarrow\; NH_4^+ \;+\; OH^-$$

In a *neutralization* reaction, the protons (H^+) from the acid combine with hydroxide ions (OH^-) from the base to produce water (H_2O). The remaining substance is a salt, which is composed of ions from the acid and base. For example, the neutralization of HCl by NaOH is written as

$$HCl \;+\; NaOH \;\longrightarrow\; NaCl \;+\; H_2O$$

If we write the ionic substances in the equation as ions, we see that the H^+ and the OH^- form water.

$$H^+ \;+\; Cl^- \;+\; Na^+ \;+\; OH^- \;\longrightarrow\; Na^+ \;+\; Cl^- \;+\; H_2O$$

$$H^+ \qquad\qquad\quad +\; OH^- \;\longrightarrow\; H_2O$$

In a complete neutralization, the amount of H^+ will be equal to the amount of OH^-.

A. pH Color Using Red Cabbage Indicator

The pH of a solution tells us whether a solution is acidic, basic, or neutral. On the pH scale, pH values below 7 are acidic, equal to 7 is neutral, and values above 7 are basic. Typically, the pH scale has values between 0 and 14.

pH scale

0 1 2 3 4 5 6 7 8 9 10 11 12 13 14

← acidic → neutral ← basic →

Many natural substances contain dyes that produce distinctive colors at different pH values. By extracting (removing) the dye from red cabbage leaves, a natural indicator can be prepared. Adding the red cabbage solution to solutions of a variety of acids and bases will produce a series of distinctive colors. When the red cabbage solution is added to a test sample, the color produced can be matched to the colors of the pH reference set to determine the pH of the sample. A pH meter can also be used to measure pH.

B. Measuring pH

The concentration (moles/liter, indicated by brackets [ ]) of H_3O^+ or OH^- can be determined from the ionization constant for water (K_w). In pure water, $[H_3O^+] = [OH^-] = 1 \times 10^{-7}$ M.

$$K_w = [H_3O^+][OH^-] = [1 \times 10^{-7}][1 \times 10^{-7}] = 1 \times 10^{-14}$$

If the $[H_3O^+]$ or $[OH^-]$ for an acid or a base is known, the other can be calculated. For example, an acid has a $[H^+] = 1 \times 10^{-4}$ M. We can find the $[OH^-]$ of the solution by solving the K_w expression for $[OH^-]$:

$$[OH^-] = \frac{1 \times 10^{-14}}{[H_3O^+]} = \frac{1 \times 10^{-14}}{1 \times 10^{-4}} = 1 \times 10^{-10} \text{ M}$$

The pH of a solution is a measure of its $[H_3O^+]$. It is defined as the negative log of the hydrogen ion concentration.

$$pH = -\log [H_3O^+]$$

Therefore, a solution with a $[H_3O^+] = 1 \times 10^{-4}$ M has a pH of 4.0, and is acidic. A solution with a $[H_3O^+] = 1 \times 10^{-11}$ M has a pH of 11.0, and is basic.

C. Effect of Buffers on pH

The pH of the blood is maintained between 7.35 and 7.45 by buffers in the body. If blood pH goes above or below that range, it can destroy the cells in the blood. *Buffers* maintain the pH of a solution by reacting with small amounts of acids or bases. Many buffers contain a weak acid and its salt. The weak acid reacts with excess base, and the anion of the salt picks up excess H^+. It is the ability of a buffer to react with added acid or base that maintains the pH of a solution. The pH of the blood is kept constant by the bicarbonate buffer, which is carbonic acid, H_2CO_3 (weak acid), and bicarbonate anion, HCO_3^- (salt). When base (OH^-) is added, it reacts with the weak acid in the buffer and produces bicarbonate ion and water:

$$H_2CO_3 + OH^- \longrightarrow HCO_3^- + H_2O$$

When acid (H^+) enters the blood, it reacts with the HCO_3^- anion and re-forms carbonic acid:

$$HCO_3^- + H^+ \longrightarrow H_2CO_3$$

In this experiment, the effect of an acid and a base on the pH of a buffer and a nonbuffer will be determined.

Lab Information

Time: $2^1/_2$ hr

Comments: Students may be asked to bring a red cabbage to class.

                     Share test tubes with your lab neighbors to prepare the pH reference solutions.

                     Tear out the report sheets and place them beside the matching procedures.

Related Topics: Acids, bases, pH, buffers

Experimental Procedures

A.   pH Color Using Red Cabbage Indicator

Materials: Red cabbage leaves, 400-mL beaker, distilled water, Bunsen burner or hot plate, 150-mL beaker, test tubes, two test tube racks, set of buffers with pH ranging from 1 to 13

Using a 150-mL beaker, obtain 50 mL of cabbage dye indicator. The indicator can be prepared by placing 5 or 6 torn leaves from red cabbage in a 400-mL beaker. Add about 150–200 mL of distilled water to cover the leaves. Heat on a hot plate or using a Bunsen burner, but do not boil. When the solution has attained a dark purple color, turn off the burner and cool.

Preparation of a pH reference set Arrange 13 test tubes in two test tube racks. You may need to combine your test tube set with your neighbor's set. (Your instructor may prepare a pH reference set for the entire class.) Pour 3–4 mL of each buffer in a separate test tube to create a set with pH values of 1–13. **Caution: Low pH values are strongly acidic; high pH values are strongly basic. Work with care.** To each test tube, add 2–3 mL of the *cooled* red cabbage solution. If you wish a deeper color, add more cabbage solution. Describe the colors of the pH solutions. *Keep this reference set for the next part of the experiment.*

B.   Measuring pH

Materials: Shell vials or test tubes, samples to test for pH (shampoo, conditioner, mouthwash, antacids, detergents, fruit juice, vinegar, cleaners, aspirin, etc.), cabbage juice indicator from part A, pH meter, calibration buffers, wash bottle, Kimwipes™

Place 3–4 mL of a sample in a shell vial (or a test tube). Add 2–3 mL of red cabbage solution. Describe the color and compare to the colors of the pH reference set. The pH of the buffer in the reference set that gives the best color match is the pH of the sample. Record. Test several samples.

pH Meter Your instructor will demonstrate the use of the pH meter and calibrate it with a known pH buffer. After you determine the pH of a sample using the red cabbage solution, take the sample to a pH meter, and record the pH. Rinse off the electrode with distilled water.

C.   Effect of Buffers on pH

Materials: Buffer with a high pH (9–11), buffer with a low pH (3–4), droppers, shell vials or test tubes, 0.1 M NaCl, 0.1 M HCl, 0.1 M NaOH, pH meter, cabbage juice indicator from part A

Effect of Adding Acid

C.1   Place 10.0 mL of one of the following solutions into a shell vial (or test tube):
      a. H_2O
      b. 0.1 M NaCl
      c. A buffer with a high pH
      d. A buffer with a low pH

Add 2–3 mL of cabbage indicator. Describe the color. Determine the pH of each sample using the pH reference set or pH meter, or both. Record.

C.2   Add 5 drops of 0.1 M HCl (acid). Stir and determine the pH. Record. Add 5 more drops of 0.1 M HCl. Record any color change in the indicator. Determine the pH. Repeat this procedure with each of the other solutions.

C.3   Determine the change in pH, if any. Identify the solutions that are buffers.

Effect of Adding Base

C.4   Place 10.0 mL of one of the following solutions into a shell vial (or test tube):
a. H_2O
b. 0.1 M NaCl
c. A buffer with a high pH
d. A buffer with a low pH

Add 2–3 mL of cabbage indicator. Describe the color. Determine the pH of each sample using the pH reference set or pH meter, or both. Record.

C.5   Add 5 drops of 0.1 M NaOH (base). Stir and determine the pH. Record. Add 5 more drops of 0.1 M NaOH. Record any color change in the indicator. Determine the pH. Repeat this procedure with the other samples listed in C.4.

C.6   Determine the change in pH, if any. Identify the solutions that are buffers.

20
Acid-Base Titration

Goals

- Prepare a sample for titration with a base.
- Set up a buret and use proper titration technique in reaching an endpoint.
- Calculate the molar concentration and percentage of acetic acid in vinegar.
- Determine the acid-absorbing capacity of a commercial antacid.

Discussion

In a neutralization reaction, the protons (H^+) from the acid combine with hydroxide ions (OH^-) from the base to produce water (H_2O). The remaining substance is a salt, which is composed of ions from the acid and base. For example, the neutralization of HCl by NaOH is written as

$$HCl + NaOH \longrightarrow NaCl + H_2O$$

If we write the ionic substances in the equation as ions, we see that the H^+ and the OH^- form water.

$$H^+ + Cl^- + Na^+ + OH^- \longrightarrow Na^+ + Cl^- + H_2O$$

$$H^+ \qquad\qquad + OH^- \longrightarrow \qquad\qquad H_2O$$

In a complete neutralization, the amount of H^+ will be equal to the amount of OH^-.

A. Acetic Acid in Vinegar

Vinegar is an aqueous solution of acetic acid, $HC_2H_3O_2$ or CH_3COOH. The amount of acetic acid in a vinegar solution can be determined by neutralizing the acid with a base, in this case NaOH. As shown in the following equation, one mole of acetic acid is neutralized by one mole of NaOH.

$$\underset{\textit{Acetic acid}}{HC_2H_3O_2} + \underset{\textit{Base}}{NaOH} \longrightarrow \underset{\textit{Salt}}{C_2H_3O_2{}^- Na^+} + H_2O$$

A *titration* involves the addition of a specific amount of base required to neutralize an acid in a sample. When all the H^+ (or H_3O^+) from the acid has been neutralized, an indicator in the sample will change color. This change in the indicator color determines the *endpoint,* which signals that the addition of the base should be stopped. The volume of base used to neutralize the acid is then determined. In this experiment, phenolphthalein is the indicator; it changes from colorless in acid to a faint but permanent pink color in base.

Calculating the molarity of acetic acid

Using the average measured volume of the NaOH, and its molarity (on the label), the moles of NaOH used can be calculated.

$$\text{Moles NaOH used} \;=\; \text{L NaOH used} \;\times\; \frac{\text{moles NaOH}}{\text{L NaOH}}$$

When an acid is completely neutralized, the moles of NaOH are equal to the moles of acetic acid ($HC_2H_3O_2$) present in the sample. This occurs because there are the same number of H^+ and OH^- ions in the reactants:

$$\text{Moles } HC_2H_3O_2 \;=\; \text{moles NaOH}$$

Using the moles of acid, the molarity of acetic acid in the 5.0-mL sample of vinegar is calculated.

$$\text{Molarity (M) } HC_2H_3O_2 \;=\; \frac{\text{moles } HC_2H_3O_2}{0.0050 \text{ L vinegar}}$$

Calculating the percent (mass/volume) of acetic acid

To calculate the percent (m/v) of $HC_2H_3O_2$ in vinegar, we convert the moles of acetic acid to grams using the molar mass of acetic acid, 60.0 g/mole.

$$\text{g } HC_2H_3O_2 \;=\; \text{moles } HC_2H_3O_2 \;\times\; \frac{60.0 \text{ g } HC_2H_3O_2}{1 \text{ mole } HC_2H_3O_2}$$

$$\text{Percent (m/v)} \;=\; \frac{\text{g } HC_2H_3O_2}{5.0 \text{ mL}} \;\times\; 100$$

B. Titration of an Antacid

Stomach acid is primarily hydrochloric acid (HCl), which has a concentration of about 0.1 M. Sometimes when a person is under stress, excess HCl may be produced, causing discomfort. An agent called an antacid is used to neutralize some of the excess stomach acid. In this experiment, the volume (mL) of 0.1 M HCl that can be absorbed by some common antacids will be determined.

Milk of magnesia, Di-Gel $\qquad Mg(OH)_2 + 2HCl \longrightarrow MgCl_2 + 2H_2O$

Tums $\qquad\qquad\qquad\qquad CaCO_3 + 2HCl \longrightarrow CaCl_2 + CO_2 + H_2O$

Lab Information

Time: 2 hr
Comments: Tear out the Lab report sheets and place them next to the matching procedures.
                Practice observing the color change for the indicators before you do the titrations.
                Carefully read the markings on the buret.
                Students may bring their own antacid samples to test.
Related topics: Acid, base, neutralization, titration, percent concentration, molarity

194

A. Acetic Acid in Vinegar

Materials: Vinegar (white), two beakers (150 and 250 mL), 250-mL Erlenmeyer flask, 10-mL graduated cylinder (or a 5-mL pipet and bulb), phenolphthalein indicator, 50-mL buret (or 25-mL buret), buret clamp, small funnel to fit buret, 0.1 M NaOH (standardized), white paper or paper towel

A.1 Obtain about 20 mL of vinegar in a small beaker. Record the brand of vinegar and the % acetic acid stated on the label. Using a 10-mL graduated cylinder or a 5.0-mL pipet, transfer 5.0 mL of vinegar to a 250-mL Erlenmeyer flask.

Using a pipet: Place the pipet bulb on the end of the pipet and squeeze the bulb to remove air. Place the tip of the pipet in the vinegar in the beaker and allow the bulb to slowly expand. (If the bulb was squeezed too much, the change in pressure will draw liquid up into the bulb.) When the liquid goes above the volume line, but not into the bulb, carefully remove the bulb and quickly place your second finger (index finger) tightly over the end of the stem. By adjusting the pressure of the index finger, lower the liquid to the etched line that marks the 5.0-mL volume and stop. Lift the pipet with its 5.0 mL of vinegar out of the beaker and let it drain into an Erlenmeyer flask. Touch the tip of the pipet to the wall of the flask to remove the rest of the vinegar. A small amount that remains in the tip has been included in the calibration of the pipet. See Figure 20.1 for use of a pipet. **Caution: If you are using a pipet, use a suction bulb to draw vinegar into a pipet. Do not pipet by mouth!**

Add about 25 mL of distilled water to increase the volume of the solution for titration. This will not affect your results. Add 2–3 drops of the phenolphthalein indicator to the solution in the flask.

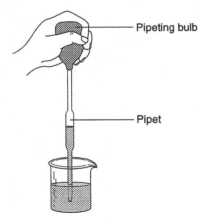

Pipeting bulb

Pipet

Figure 20.1 Using a bulb to draw a liquid into a pipet

A.2 Obtain a 50-mL or 25-mL buret and place it in a buret clamp or butterfly clamp as shown in Figure 20.2. Using a 250-mL beaker, obtain about 100 mL of 0.1 M NaOH solution. (If you are using a 25.0-mL buret, use a 0.2 M NaOH solution.) Record the molarity (M) of the NaOH solution that is stated on the label of the reagent bottle. Rinse the buret with two 5-mL portions of the NaOH solution. Discard the NaOH washings.

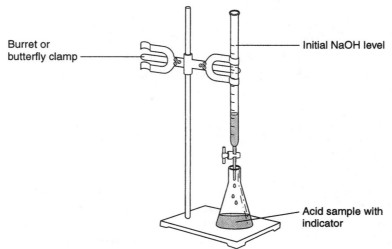

Figure 20.2 Buret setup for acid–base titration

A.3 Observe the markings on the buret. The top is marked 0.0 mL, and 50.0 mL (or 25.0 mL) is marked at the bottom. Place a small funnel in the top of the buret and carefully pour NaOH into the funnel. Pour slowly as the NaOH fills the buret. Lift the funnel and allow the NaOH to go above the top line (0.0 mL). Slowly open the stopcock and drain NaOH into a waste beaker until the meniscus is at the 0.00 mL line or below. The buret tip should be full of NaOH solution, and free of bubbles. Record the initial buret reading of NaOH.

A.4 Place the flask containing the vinegar solution under the buret on a piece of white paper. (Be sure you added indicator.) Begin to add NaOH to the solution by opening and closing the stopcock with your left hand (if you are right-handed). Swirl the flask with your right hand to mix the acid and the base. At first, the pink color produced by the reaction will disappear quickly. As you near the endpoint, the pink color will be more persistent and disappear slowly. *Slow down* the addition of the NaOH to drops at this time. Soon, one drop of NaOH will give a faint, permanent pink color to the sample. *Stop adding NaOH.* You have reached the endpoint of the titration. See Figure 20.3.

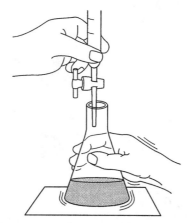

Figure 20.3 During a titration, the solution in the flask is swirled as NaOH is added to the acid sample.

At the endpoint, record the final buret reading of the NaOH. Fill the buret again. Repeat the titration with new samples of the same brand of vinegar. *Be sure to add water and indicator to each new sample of vinegar.*

Calculations

A.5 Calculate the volume of NaOH used to neutralize the vinegar sample(s).

$$\text{Volume} = \text{Final volume (NaOH)} - \text{initial volume (NaOH)}$$

After the titration of two or three samples of vinegar, calculate the average volume of NaOH used. Total the volumes of NaOH used, and divide by the number of samples you used.

$$\text{mL (average)} \quad \frac{\text{Volume (1)} + \text{Volume (2)} + \text{Volume (3)}}{3}$$

Calculating the molarity of acetic acid

A.6 Convert the volume (average) of NaOH used to a volume in liters (L).

$$L = \text{Average volume (mL) NaOH} \times \frac{1\ L}{1000\ mL}$$

A.7 Calculate the moles of NaOH using the volume (L) and molarity of the NaOH.

$$\text{Moles NaOH used} = L\ \text{NaOH used} \times \frac{\text{moles NaOH}}{1\ L\ \text{NaOH}}$$

A.8 Record the moles of acid present in the vinegar, which are equal to the moles of NaOH used.

$$\text{Moles } HC_2H_3O_2 = \text{moles NaOH used}$$

A.9 Calculate the molarity (M) of the acetic acid ($HC_2H_3O_2$) in the vinegar sample.

$$\text{Molarity (M) } HC_2H_3O_2 = \frac{\text{moles } HC_2H_3O_2}{0.0050\ L\ \text{vinegar}}$$

Calculating the percent (mass/volume) of acetic acid

A.10 Using the number of moles of acetic acid, calculate the number of grams of acetic acid in the vinegar sample. The molar mass of acetic acid is 60.0 g/mole.

$$g\ HC_2H_3O_2 = \text{moles } HC_2H_3O_2 \times \frac{60.0\ g\ HC_2H_3O_2}{1\ \text{mole } HC_2H_3O_2}$$

A.11 Calculate the percent (mass/volume) acetic acid in the vinegar. For an original volume of 5.0 mL of vinegar, the percent is calculated as follows:

$$\text{Percent (m/v)} = \frac{g\ HC_2H_3O_2}{5.0\ mL} \times 100$$

B. Titration of an Antacid

Materials: Antacid products, two small beakers (100- and 150-mL), mortar and pestle, 250-mL Erlenmeyer flask, methyl orange, graduated cylinder, 0.10 M HCl (standardized), 0.10 M NaOH (standardized), buret, small funnel, buret clamp, white paper or paper towel

B.1 Record the name of the antacid and its active ingredients listed on the label.

B.2 Use a mortar and pestle to crush an antacid tablet. Weigh a 250-mL Erlenmeyer flask. Record its mass. Do not round off. Transfer some of the crushed tablet (up to 0.5 g) to the flask. Weigh the flask and the crushed antacid. Record.

B.3 Add 50.0 mL of HCl to the flask. Record the molarity of the HCl as stated on the reagent bottle. Swirl to dissolve the antacid. The solution may be cloudy because of the starches in the tablet. This will not affect the titration. Add 5 drops of methyl orange. The solution should be red. If the solution is yellow, add more HCl in 10.0-mL portions until the solution is red. Record the total volume of HCl that you add.

B.4 Set up a buret using 0.1 M NaOH. Record the molarity of NaOH as stated on the reagent bottle.

B.5 Titrate the sample using 0.1 M NaOH until a yellow color forms. This titration neutralizes the amount of HCl that was not neutralized by the antacid. Record the initial and final volumes of NaOH used.

Calculations

B.6 Calculate the mass of the antacid used.

B.7 Calculate the volume of NaOH used in the titration.

$$mL\ NaOH\ used\ =\ final\ level\ NaOH\ -\ initial\ level\ NaOH$$

B.8 State the volume of excess HCl. This is equal to the volume of NaOH because the molarity of the NaOH matched the molarity of the HCl.

B.9 Calculate the volume of HCl that was neutralized by the antacid. This is the difference between the total volume of HCl added to the antacid minus the volume of the excess HCl.

B.10 Because stomach acid is about the same molarity as 0.1 M HCl, we can state the neutralizing power of the antacid as mL of stomach acid per gram of antacid.

$$\frac{mL\ 0.1\ M\ HCl\ neutralized}{grams\ antacid}\ =\ \frac{mL\ stomach\ acid\ neutralized}{1\ g\ antacid}$$

Record values (mL stomach acid/1 g antacid) obtained in the class for other antacids.

B.11 Write neutralization equations for the bases in the antacid products.

Report Sheet - Lab 20

Date _____   Name _____

Section _____   Team _____

Instructor _____

Pre-Lab Study Questions

1. What is neutralization?

2. Write an equation for the neutralization of H_2SO_4 by KOH.

3. What is a titration?

4. What is the function of an indicator in a titration?

5. What is the purpose of taking an antacid?

Report Sheet - Lab 20

A. Acetic Acid in Vinegar

A.1 Brand _____ Volume <u>5.0 mL</u> (% on label) _____%

A.2 Molarity (M) of NaOH (stated on label) _____M

		Trial 1	Trial 2	Trial 3
A.3	Initial NaOH level in buret			
A.4	Final NaOH level in buret			
A.5	Volume (mL) of NaOH used			
	Average volume (mL)			
A.6	Average volume in liters (L)			

A.7 Moles of NaOH used in titration   _____ mole NaOH
     (Show calculations.)

A.8 Moles of $HC_2H_3O_2$ neutralized by NaOH   _____ mole $HC_2H_3O_2$

A.9 Molarity of $HC_2H_3O_2$ _____ M $HC_2H_3O_2$
     (Show calculations.)

A.10 Grams $HC_2H_3O_2$ _____ g $HC_2H_3O_2$
     (Show calculations.)

A.11 Percent (m/v) $HC_2H_3O_2$ in vinegar   _____% $HC_2H_3O_2$
     (Show calculations.)

Questions and Problems

Q.1 How many grams of $Mg(OH)_2$ will be needed to neutralize 25 mL of stomach acid if stomach acid is 0.10 M HCl?

Q.2 How many mL of a 0.10 M NaOH solution are needed to neutralize 15 mL of 0.20 M H_3PO_4 solution?

Report Sheet - Lab 20

B. Titration of an Antacid

		Antacid 1	Antacid 2	Antacid 3
B.1	Brand of antacid			
	Active ingredient(s)			
B.2	Mass of flask and antacid			
	Mass of flask			
B.3	Molarity of HCl			
	Total volume (mL) HCl			
B.4	Molarity of NaOH			
B.5	Final volume (mL) NaOH			
	Initial volume (mL) NaOH			
Calculations				
B.6	Mass of antacid			
B.7	Volume of NaOH used			
B.8	Volume of excess HCl			
B.9	Volume of HCl neutralized by antacid			
B.10	mL stomach acid 1 g antacid			

Report Sheet - Lab 20

B.11 Write the neutralization equation that would take place in the stomach with two different bases used in the antacid products:

1._____

2._____

Properties of Organic Compounds

Goals

- Observe chemical and physical properties of organic and inorganic compounds.
- Identity functional groups in three-dimensional models.

Discussion

A. Color, Odor, and Physical State

Organic compounds are made of carbon and hydrogen, and sometimes oxygen and nitrogen. Of all the elements, only carbon atoms bond to many more carbon atoms, a unique ability that gives rise to many more organic compounds than all the inorganic compounds known today. The covalent bonds in organic compounds and the ionic bonds in inorganic compounds account for several of the differences we will observe in their physical and chemical properties. See Table 21.1.

Table 21.1 *Comparing Some Properties of Organic and Inorganic Compounds*

Organic Compounds	Inorganic Compounds
Covalent bonds	Ionic or polar bonds
Soluble in nonpolar solvents, not water	Soluble in water
Low melting and boiling points	High melting and boiling points
Strong, distinct odors	Usually no odor
Poor or nonconductors of electricity	Good conductors of electricity
Flammable	Not flammable

B. Solubility

Typically, inorganic compounds that are ionic are soluble in water, a polar compound, but organic compounds are nonpolar and thus are not soluble in water. However, organic compounds are soluble in organic solvents because they are both nonpolar. A general rule for solubility is that "like dissolves like."

C. Combustion

Many organic compounds react with oxygen, a reaction called *combustion,* to form carbon dioxide and water. Combustion is the reaction that occurs when gasoline burns with oxygen in the engine of a car or when natural gas, methane, burns in a heater or stove. In a combustion reaction, heat is given off; the reaction is exothermic. Equations for the combustion of methane and propane are written as follows:

$$CH_4(g) \ + \ 2O_2(g) \ \longrightarrow \ CO_2(g) \ + \ 2H_2O(g) \ + \ heat$$
methane

$$C_3H_8(g) \ + \ 5O_2(g) \ \longrightarrow \ 3CO_2(g) \ + \ 4H_2O(g) \ + \ heat$$
propane

D. Functional Groups

Although there are millions of organic compounds, they can be classified according to organic families. Each family contains a characteristic structural feature called a functional group, which is a certain atom or group of atoms that give similar physical and chemical properties to that family. Because the organic compounds in a family contain the same functional group, they undergo the same types of chemical reactions. In this lab, we will take a look at some of the common functional groups that allow us to classify organic compounds according to the structure.

Alkanes, alkenes, and alkynes are hydrocarbons that consist of only carbon and hydrogen atoms. Alkanes contain carbon-carbon single bonds, whereas alkenes contain one or more carbon-carbon double bonds, and alkynes contain a carbon-carbon triple bond. To write a condensed structural formula, the hydrogen atoms attached to each carbon are written adjacent to the symbol C for carbon. Thus, a CH_3- is the abbreviation for a carbon attached to three hydrogen atoms, whereas $-CH_2-$ shows a carbon attached to two hydrogen atoms.

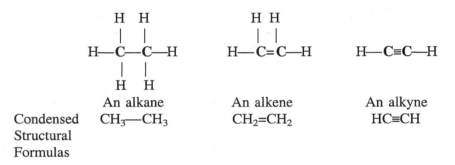

Condensed Structural Formulas

Alcohols and ethers contain an oxygen atom. Alcohols have a *hydroxyl group*, which is an –OH group, bonded to a carbon atom. In an ether, the oxygen atom is bonded to two carbon atoms.

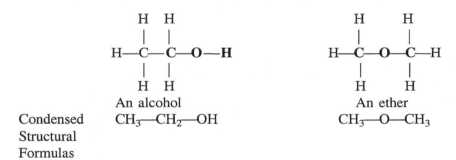

Condensed Structural Formulas

Aldehydes and ketones, contain a carbonyl group, which is a carbon-oxygen double bond (C=O). In an aldehyde, the carbon bonds to at least one hydrogen atom. In a ketone, the carbon in the carbonyl group bonds to two other carbon atoms.

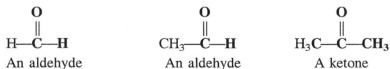

Carboxylic acids and esters contain the functional group is the carboxyl group, which is a combination of a carbonyl and hydroxyl group. In a carboxylic acid, the oxygen is bonded to a hydrogen atom, whereas in an ester the oxygen is bonded to a carbon and not to hydrogen.

$$\overset{\overset{\displaystyle O}{\overset{\displaystyle \|}{}}}{CH_3-C-O-H} \qquad \text{or} \qquad CH_3COOH \qquad \text{or} \qquad CH_3CO_2H$$

A carboxylic acid

$$\overset{\overset{\displaystyle O}{\overset{\displaystyle \|}{}}}{CH_3-C-O-CH_3} \qquad \text{or} \qquad CH_3COOCH_3 \qquad \text{or} \qquad CH_3CO_2CH_3$$

An ester

Amines contain a nitrogen atom because they are derivatives of ammonia, NH_3. In an amine, one or more carbon groups replace the hydrogen atoms in ammonia. Amines are classified as primary, secondary, or tertiary according to the number of carbon groups bonded to the nitrogen atoms.

$$NH_3 \qquad CH_3-NH_2 \qquad \overset{\displaystyle CH_3-NH}{\underset{\displaystyle CH_3}{|}} \qquad \overset{\displaystyle CH_3-N-CH_3}{\underset{\displaystyle CH_3}{|}}$$

Ammonia   A primary   A secondary   A tertiary
               amine (1°)   amine (2°)   amine (3°)

Lab Information

Time: 2 hr
Comments: Tear out the report sheets and place them next to the matching procedures. *Organic compounds are extremely flammable! Use of the Bunsen burner is prohibited.*
Related Topics: Organic compounds, hydrocarbons, solubility, combustion, complete structural formula, functional groups

Experimental Procedures   Wear your safety goggles!

A. Color, Odor, and Physical State *(This may be a lab display.)*

Materials: Test tubes (6), test tube rack, spatulas, NaCl(*s*), KI(*s*), toluene, benzoic acid, cyclohexane, water, chemistry handbook

Place each substance into a separate test tube: a few crystals of NaCl, KI, and benzoic acid, and 10 drops each of cyclohexane, toluene, and water. Or if a display is available, observe the samples in a test tube rack in the hood. Record the formula, physical state (solid, liquid, or gas), and odor of each one. To check for odor, first take a breath and hold it while you gently fan the air above the test tube toward you. Look up the melting point of each compound using a chemistry handbook. Record. State the types of bonds in each as ionic or covalent. Identify each as an organic or inorganic compound.

B. **Solubility** *(This may be a demonstration or lab display.)*

Materials: Test tubes, spatulas, NaCl(*s*), toluene, cyclohexane

Work in the hood: Be sure to work with the compounds such as cyclohexane in ventilation hoods, and then dispose of them in the proper waste containers. Place 10 drops of cyclohexane and 10 drops of water in a test tube. Record your observations. Identify the upper layer and the lower layer.

Place a few crystals of NaCl in one test tube and 10 drops of toluene in another test tube. To each sample, add 15 drops of cyclohexane, a nonpolar solvent. Shake gently or tap the bottom of the test tube to mix. Record whether each substance is soluble (S) or insoluble (I) in cyclohexane.

Repeat the experiment with the two substances, but this time add 15 drops of water, a polar solvent. Record whether each substance is soluble (S) or insoluble (I) in water. Identify each substance as an organic or inorganic compound.

Dispose of organic substances in the proper waste container.

C. **Combustion** *(This may be a demonstration by your instructor.)*

Materials: 2 evaporating dishes, spatulas, wood splints, NaCl(*s*), cyclohexane

Work in the hood: Place a small amount (pea-size) of NaCl in an evaporating dish set in an iron ring. Ignite a splint and hold the flame to the NaCl. Record whether the substance burns. Repeat the experiment using 5 drops of cyclohexane instead of NaCl. If the substance burns, note the color of the flame. Identify each as an organic or inorganic compound.

D. **Functional Groups**

Materials: An organic model kit or prepared models of organic compounds

D.1 Observe models of organic compounds listed in the table. Or using an organic model kit, construct ball-and-stick models of each of the compounds. Place wooden dowels or springs in all the holes in the carbon atom (black) and attach hydrogen (yellow) atoms, oxygen atoms (red), or nitrogen atoms (blue) as required to complete the functional group for each. Draw a full structural formula showing all the bonds to each carbon atom.

D.2 Circle the functional group in each structure. Classify the organic compound according to the functional group

Report Sheet - Lab 21

Date _____   Name _____

Section _____   Team _____

Instructor _____

Pre-Lab Study Questions

1. Would you expect an organic compound to be soluble in water? Why?

2. Which is more flammable: an organic or inorganic compound?

A. Color, Odor, and Physical State

Name	Formula	Physical State	Odor	Melting Point	Type of Bonds?	Organic or Inorganic?
Sodium chloride						
Cyclohexane	C_6H_{12}					
Potassium iodide						
Benzoic acid	$C_7H_6O_2$					
Toluene	C_7H_8					
Water						

B. Solubility

In the mixture, water is the _____ layer and cyclohexane is the _____ layer.

Solute	Solubility in Cyclohexane	Solubility in Water	Organic or Inorganic?
NaCl			
Toluene			

Report Sheet - Lab 21

C. Combustion

Compound	Flammable (Color of Flame)	Not Flammable	Organic or Inorganic?
NaCl			
Cyclohexane			

From your observations of the chemical and physical properties of alkanes as organic compounds, complete the following table:

Property	Organic Compounds	Inorganic Compounds
Elements		
Bonding		
Melting points		
Strong odors		
Flammability		
Solubility		

Questions and Problems

Q.1 Describe three properties you can use to distinguish between organic and inorganic compounds.

Q.2 A white solid has no odor, is soluble in water, and is not flammable. Would you expect it to be an organic or an inorganic substance? Why?

Q.3 A clear liquid with a gasoline-like odor forms a layer when added to water. Would you expect it to be an organic or an inorganic substance? Why?

Report Sheet - Lab 21

D. Functional Groups

Compound	D.1 Full Structural Formula	D.2 Organic Family
CH_3—OH		
CH_3—CH_2—CH_3		
CH_2=CH_2		
CH_3—O—CH_3		
CH_3—NH_2		
$\begin{matrix} & O \\ & \parallel \\ CH_3 &-C-OH \end{matrix}$		
$\begin{matrix} & O \\ & \parallel \\ CH_3 &-C-CH_3 \end{matrix}$		
$\begin{matrix} & H \\ & \mid \\ CH_3 &-N-CH_3 \end{matrix}$		

Report Sheet - Lab 21

Questions and Problems

Q.4 Classify the following organic compounds according to their functional groups:

a. _____ CH_3—CH_2—CH=CH—CH_3

b. _____
$$CH_3—CH_2—\overset{\overset{H}{|}}{N}—CH_2—CH_3$$

c. _____ CH_3—CH_2—O—CH_3

d. _____
$$CH_3—\overset{\overset{O}{\|}}{C}—CH_2—CH_3$$

e. _____
$$CH_3—CH_2—CH_2—\overset{\overset{O}{\|}}{C}—OH$$

f. _____
$$CH_3—CH_2—CH_2—\overset{\overset{O}{\|}}{C}—O—CH_3$$

22

Structures of Alkanes

Goals

- Draw formulas for alkanes from their three-dimensional models.
- Write the names of alkanes from their structural formulas.
- Construct models of isomers of alkanes.
- Write the structural formulas for cycloalkanes and haloalkanes.

Discussion

A. Structures of Alkanes

The saturated hydrocarbons represent a group of organic compounds composed of carbon and hydrogen. Alkanes and cycloalkanes are called *saturated* hydrocarbons because their carbon atoms are connected by only single bonds. In each type of alkane, each carbon atom has four valence electrons and must always have four single bonds.

To learn more about the three-dimensional structure of organic compounds, it is helpful to build models using a ball-and-stick model kit. In the kit are wooden (or plastic) balls, which represent the typical elements in organic compounds. Each wooden atom has the correct number of holes drilled for bonds that attach to other atoms. See Table 22.1.

Table 22.1 *Elements and Bonds Represented in the Organic Model Kit*

Color	Element	Number of Bonds
Black	carbon	4
Yellow	hydrogen	1
Red	oxygen	2
Green	chlorine	1
Orange	bromine	1
Purple	iodine	1
Blue	nitrogen	3
Bonds		
Sticks, springs		

The first model to build is methane, CH_4, a hydrocarbon consisting of one carbon atom and four hydrogen atoms. The model of methane shows the three-dimensional shape, a tetrahedron, around a carbon atom.

Three-dimensional structure   Complete structural formula   Condensed structural formula

211

To represent this model on paper, its shape is flattened, and the carbon atom is shown attached to four hydrogen atoms. This type of formula is called a *complete structural formula.* However, it is more convenient to use a shortened version called a *condensed structural formula.* To write a condensed formula, the hydrogen atoms are grouped with their carbon atom. The number of hydrogen atoms is written as a subscript. The complete structural formula and the condensed structural formula for C_2H_6 are shown below:

H   H
| |
H—C—C—H CH_3—CH_3 or CH_3CH_3
| |
H   H

Complete structural formula   Condensed structural formula

Names of Alkanes

The names of alkanes all end with *-ane*. The names of organic compounds are based on the names of the alkane family. See Table 22.2.

Table 22.2   *Names and Formulas of the First Ten Alkanes*

Name	Formula	Name	Formula
Methane	CH_4	Hexane	$CH_3CH_2CH_2CH_2CH_2CH_3$
Ethane	CH_3CH_3	Heptane	$CH_3CH_2CH_2CH_2CH_2CH_2CH_3$
Propane	$CH_3CH_2CH_3$	Octane	$CH_3CH_2CH_2CH_2CH_2CH_2CH_2CH_3$
Butane	$CH_3CH_2CH_2CH_3$	Nonane	$CH_3CH_2CH_2CH_2CH_2CH_2CH_2CH_2CH_3$
Pentane	$CH_3CH_2CH_2CH_2CH_3$	Decane	$CH_3CH_2CH_2CH_2CH_2CH_2CH_2CH_2CH_2CH_3$

B.   Constitutional Isomers

Constitutional Isomers are present when a molecular formula can represent two or more different structural (or condensed) formulas. One structure cannot be converted to the other without breaking and forming new bonds. The isomers have different physical and chemical properties. One of the reasons for the vast array of organic compounds is the phenomenon of isomerism.

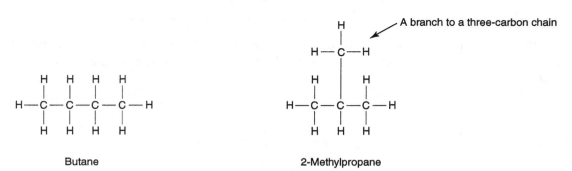

Isomers of C_4H_{10}

Butane   2-Methylpropane

C. Cycloalkanes

In a cycloalkane, an alkane has a cyclic or ring structure. There are no end carbon atoms. The structural formula of a cycloalkane indicates all of the carbon and hydrogen atoms. The condensed formula groups the hydrogen atoms with each of the carbon atoms. Another type of notation called the *geometric* structure is often used to depict a cycloalkane by showing only the bonds that outline the geometric shape of the compound. For example, the geometric shape of cyclopropane is a triangle, and the geometric shape of cyclobutane is a square. Examples of the various structural formulas for cyclobutane are shown below.

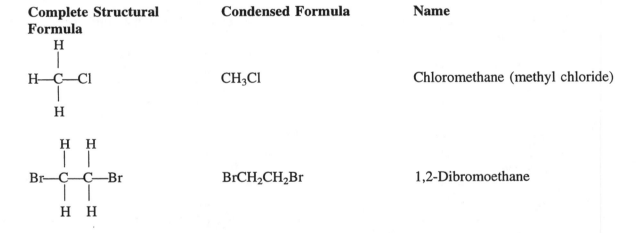

| Complete structural formula | Condensed formula | Geometric formula |

D. Haloalkanes

In a haloalkane, a halogen atom such as chlorine (Cl) or bromine (Br) replaces a hydrogen atom of an alkane or a cycloalkane.

Complete Structural Formula	Condensed Formula	Name
(structure: $H-\overset{H}{\underset{H}{C}}-Cl$)	CH_3Cl	Chloromethane (methyl chloride)
(structure: $Br-\overset{H}{\underset{H}{C}}-\overset{H}{\underset{H}{C}}-Br$)	$BrCH_2CH_2Br$	1,2-Dibromoethane

Lab Information

Time: 2 hr
Comments: Tear out the report sheets and place them next to the matching procedures.
Related Topics: Alkane, cycloalkane, haloalkane, complete structural formula, condensed structural formula, constitutional isomers, naming alkanes

Experimental Procedures

A. Structures of Alkanes

Wear your safety goggles!

Materials: Organic model kit

A.1 Using an organic model kit, construct a ball-and-stick model of a molecule of methane, CH_4. Place wooden dowels in all the holes in the carbon atom (black). Attach hydrogen (yellow) atoms to each. Draw the three-dimensional (tetrahedral) shape of methane. Write the complete structural formula and the condensed structural formula of methane.

A.2 Make a model of ethane, C_2H_6. Observe that the tetrahedral shape is maintained for each carbon atom in the structure. Write the complete structural and condensed structural formulas for ethane.

A.3 Make a model of propane, C_3H_8. Write the complete structural and condensed structural formulas for propane.

B. Constitutional Isomers

Materials: Organic model kit, chemistry handbook

B.1 The molecular formula of butane is C_4H_{10}. Construct a model of butane by connecting four carbon atoms in a chain. Draw its complete and condensed structural formulas.

B.2 Make an isomer of C_4H_{10}. Remove an end -CH_3 group and attach it to the center carbon atom. Complete the end of the chain with a hydrogen atom. Write its complete and condensed structural formulas.

B.3 Obtain a chemistry handbook. For each isomer, find the molar mass, melting point, boiling point, and density.

B.4 Make models of the three isomers of C_5H_{12}. Make the continuous-chain isomer first. Draw the complete structural and condensed structural formulas for each. Name each isomer.

B.5 Obtain a chemistry handbook, and find the molar mass, melting point, boiling point, and density of each structural isomer.

C. Cycloalkanes

Materials: Organic model kit

C.1 Use the springs in the model kits to make a model of the cycloalkane with three carbon atoms. Write the complete structural and condensed structural formulas. Draw the geometric formula and name this compound.

C.2 Use the springs in the model kits to make models of a cycloalkane with four carbon atoms, and one with five carbon atoms. Draw the complete structural and condensed structural formulas. Draw the geometric formula and give the name for each.

D. Haloalkanes

Materials: Organic model kit

D.1 Make a model of chloromethane using a green wooden ball for chlorine. Draw the complete structural and condensed structural formulas.

D.2 Make a model of 1,2-dibromoethane using orange wooden balls for the bromine atoms. Draw the complete structural and condensed structural formulas.

D.3 Make a model of 2-iodopropane using a violet wooden ball for iodine. Draw the complete structural and condensed structural formulas.

D.4 Prepare models of four isomers of dichloropropane. Draw the complete structural and condensed structural formulas. Name each isomer.

Report Sheet - Lab 22

Date _____   Name _____

Section _____   Team _____

Instructor _____

Pre-Lab Study Questions

1. What elements are present in alkanes?

2. How does a complete structural formula differ from a condensed structural formula?

3. If isomers of an alkane have the same molecular formula, how do they differ?

A.   Structures of Alkanes

A.1 Structure of methane		
Tetrahedral shape	Complete structural formula	Condensed structural formula

A.2 Structure of ethane	
Complete structural formula	Condensed structural formula

A.3 Structure of propane	
Complete structural formula	Condensed structural formula

Report Sheet - Lab 22

Questions and Problems

Q.1 Write the correct name of the following alkanes:

a. $CH_3CH_2CH_3$ _____

b.
$$CH_3$$
$$|$$
$$CH_3CH_2CHCHCH_3$$ _____
$$|$$
$$CH_3$$

c.
$$CH_3 \quad CH_3$$
$$| \qquad |$$
$$CH_3{-}CH{-}CH{-}CH_3$$ _____

Q.2 Write the condensed formulas for the following:

a. hexane   b. 2,3-dimethylpentane

B. Constitutional Isomers

B.1 Butane C_4H_{10}	
Complete structural formula	Condensed structural formula
B.2 2-Methylpropane	
Complete structural formula	Condensed structural formula

Report Sheet - Lab 22

B.3 Physical Properties of Isomers of C_4H_{10}				
Isomer	**Molar Mass**	**Melting Point**	**Boiling Point**	**Density**
Butane				
2-Methylpropane (isobutane)				

Questions and Problems

Q.3   In B.3, what physical property is identical for the two isomers of C_4H_{10}?

Q.4   What physical properties are different for the isomers? Explain.

Report Sheet - Lab 22

B.4 Isomers of C_5H_{12}

Complete structural formula	Condensed structural formula

Name:

Complete structural formula	Condensed structural formula

Name:

Complete structural formula	Condensed structural formula

Name:

B.5 Physical Properties of Isomers of C_5H_{12}

Isomer	Molar Mass	Melting Point	Boiling Point	Density
Pentane				
2-Methylbutane				
2,2-Dimethylpropane				

Report Sheet - Lab 22

Questions and Problems

Q.5   Write the condensed formulas for the five isomers of C_6H_{14}.

C.   Cycloalkanes

Complete Structural Formula	Condensed Structural Formula	Geometric Formula
C.1 Three carbon atoms		
Name:		
C.2 Four carbon atoms		
Name:		
Five carbon atoms		
Name:		

Report Sheet - Lab 22

D. Haloalkanes

Complete Structural Formula	Condensed Structural Formula
D.1 **Chloromethane**	
D.2 **1,2-Dibromoethane**	
D.3 **2-Iodopropane**	

Report Sheet - Lab 22

D.4 Four isomers of dichloropropane

Complete Structural Formula	Condensed Structural Formula
Name:	
Name:	
Name:	
Name:	

Report Sheet - Lab 22

Questions and Problems

Q.6 Write the condensed structural formulas and names for all the constitutional isomers with the formula C_4H_9Cl.

23

Reactions of Hydrocarbons

Goals

- Observe the reactions of hydrocarbons with oxygen, bromine, and potassium permanganate.
- Use chemical tests to distinguish alkanes from alkenes.
- Draw the products of combustion, addition, and/or substitution reactions of alkanes and alkenes.

Discussion

A. Types of Hydrocarbons

Alkanes are saturated hydrocarbons, containing single bonds between carbon atoms. The alkenes and alkynes are unsaturated hydrocarbons, containing double or triple bonds. The double or triple bond, which is unsaturated, is a very reactive site in an alkene or alkyne. The aromatic compounds are hydrocarbons with a benzene ring. See Table 23.1.

Table 23.1 *Bonding Characteristics of Hydrocarbon Families*

Alkane	Alkene	Alkyne	Aromatic
Single bond	Double bond	Triple bond	Benzene ring
H H | | H—C—C—H | | H H	H H | | H—C=C—H	H—C≡C—H	(benzene ring structure)
Ethane	Ethene (ethylene)	Ethyne (acetylene)	Benzene

B. Combustion *(This may be an instructor demonstration.)*

When a compound burns in the presence of oxygen, the reaction is called combustion. This is the reaction that occurs when the methane gas in a Bunsen burner, a gas range, or a heater is ignited. The products of combustion are carbon dioxide (CO_2) and water (H_2O).

$$CH_4 + 2O_2(g) \xrightarrow{\text{Heat}} CO_2(g) + 2H_2O(g)$$

Methane

C. Bromine Test

When bromine (Br_2) reacts with an alkene, the dark red color of the Br_2 disappears quickly as the atoms of bromine bond with the carbon atoms in the double bond. If the red color disappears rapidly, we know the compound contains an unsaturated site.

$$CH_3\text{—}CH\text{=}CH\text{—}CH_3 \;+\; Br_2 \longrightarrow CH_3\text{—}\underset{|}{CH}\text{—}\underset{|}{CH}\text{—}CH_3$$

with Br, Br labels above the two central carbons.

Colorless   *Red*   *Colorless*

Bromine reacts with alkanes by replacing an H with a Br. However, the reaction is slow and requires light. Then the red bromine color persists for several minutes before it fades. Aromatic compounds (benzene ring) are not reactive with bromine. However, the methyl in toluene can react, but slowly.

$$CH_3CH_3 \;+\; Br_2 \xrightarrow[\text{(slowly)}]{\text{Light}} CH_3CH_2\text{—}Br \;+\; HBr(g)$$

Colorless   *Red*   *Colorless*   *Pungent odor*

benzene ring $+\; Br_2 \longrightarrow$ No reaction

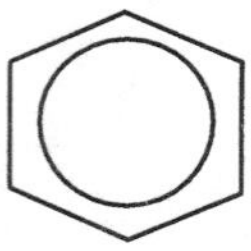

D. Potassium Permanganate ($KMnO_4$) Test

In this test, potassium permanganate ($KMnO_4$) reacts with alkenes, but not with alkanes or aromatic compounds. In the reaction, the purple color of $KMnO_4$ changes to the muddy brown of manganese dioxide (MnO_2). The product is a diol.

$$\text{Alkene} \;+\; KMnO_4 \longrightarrow \text{Diol} \;+\; MnO_2(s)$$

              Purple   *Brown solid*

E. Identification of Unknown

Use the test results to identify your unknown as one of the compounds used in this experiment.

Lab Information

Time:	2 hr
Comments:	Tear out the report sheets and place them beside the matching procedures.
	Caution: Hydrocarbons are flammable. Use very small amounts. Do not use any burners during these labs. Avoid touching the chemicals. Dispose of organic wastes in proper containers.

Related topics: Saturated, unsaturated, and aromatic hydrocarbons, haloalkanes, combustion, addition reactions

Experimental Procedures

Wear your safety goggles!

A. Types of Hydrocarbons

Materials: Organic model kit

Using the organic model kits, make models of ethene (ethylene), propene, cyclobutene, cis-2-butene, and ethyne (acetylene). Use the springs in the kit to form double bonds, triple bonds, and rings. A cis isomer of an alkene has the carbon groups attached on the same side of the double bond; a trans isomer has the groups attached on opposite sides. Draw their condensed structural formulas.

B. Combustion *(This may be an instructor demonstration.)*

Materials: Evaporating dish, wooden splints, matches, cyclohexane, cyclohexene, toluene, unknowns

B.1 *Working in the hood*, place 5 drops of cyclohexane on an evaporating dish. Using a lighted splint, *carefully* ignite the sample. Repeat the combustion test with 5 drops of cyclohexene, toluene, unknowns. Observe the flame and type of smoke produced by each. Record your observations.

B.2 Write the equations for the combustion reactions of cyclohexane (C_6H_{12}), cyclohexene (C_6H_{10}), and toluene (C_7H_8).

C. Bromine Test *(This may be an instructor demonstration.)*

Materials: 4 test tubes, test tube rack, dropper bottle of 1% bromine solution (in methylene chloride), cyclohexane, cyclohexene, toluene, unknowns

Caution: Work in the hood. The fumes of Br_2 can irritate the throat and sinuses. If bromine is spilled on the skin, flood with water for 10 minutes.

C.1 Place 15 drops of each hydrocarbon in a separate dry test tube. Label. Carefully add 3–4 drops of the bromine solution to each. Observe whether the red color disappears immediately or not. Hold the test tubes containing cyclohexane and toluene in a window with direct light. Observe whether the red color disappears, and if the odor of HBr given off can be detected.

C.2 Draw the condensed structural formula of each reactant, and of its products if a reaction occurred. If no reaction occurs, write "no reaction" (NR).

D. Potassium Permanganate ($KMnO_4$) Test

Materials: 4 test tubes, test tube rack, 1% $KMnO_4$, cyclohexane, cyclohexene, toluene, unknowns

Place 5 drops of each hydrocarbon in a separate test tube. Add 15 drops of 1% $KMnO_4$ solution. *Caution: $KMnO_4$ stains the skin*. A positive test for an unsaturated compound is a change in color from purple to brown in 60 seconds or less. Record your observations.

E. Identification of Unknown

From your test results, identify your unknown as a saturated (alkane) or unsaturated (alkene) hydrocarbon. Give your reasoning. If the names of the possible compounds are known, write their names and condensed structural formulas.

Report Sheet - Lab 23

Date _____   Name _____

Section _____   Team _____

Instructor _____

Pre-Lab Study Questions

1. What changes in color occur when bromine or $KMnO_4$ reacts with an alkene?

2. What are the products of combustion of an organic compound?

3. Why is the reaction of ethene with bromine called an addition reaction?

A.   Types of Hydrocarbons

Models of Unsaturated Hydrocarbons

Name	Condensed Structural Formula
Ethene	
Propene	
Cyclobutene	
cis-2-Butene	
Ethyne (acetylene)	

Report Sheet - Lab 23

Questions and Problems

Q.1 Write the names of the following compounds:

a. CH_3—$CH=CH_2$ _____

b. $CH_3CH=CH$—CH_3 _____

c. $CH_2=C$—$CH_2CH_2CH_3$ with CH_3 branch _____

d. _____

e. $HC\equiv CH$ _____

f. (benzene ring) _____

Q.2 Draw the structural feature that is characteristic of the following types of hydrocarbons:

Alkane	Alkene	Alkyne	Aromatic

Report Sheet - Lab 23

B.   Combustion

Hydrocarbon	B.1 Observations of Combustion	B.2 Balanced Equation for Combustion
Cyclohexane (C_6H_{12})		
Cyclohexene (C_6H_{10})		
Toluene (C_7H_8)		
Unknown		

C.   Bromine Test and D. Potassium Permanganate (KMnO$_4$) Test

	C.1 Bromine Test Observations	D. KMnO$_4$ Test Observations
Cyclohexane (C_6H_{12})		
Cyclohexene (C_6H_{10})		
Toluene (C_7H_8)		
Unknown		

Report Sheet - Lab 23

C.2

	Cyclohexane	Cyclohexene	Toluene
Condensed structural formula			
Product with bromine (if reaction)			

Questions and Problems

Q.3 Complete and balance the following reactions:

a. CH_3—$CH=CH$—CH_3 + Br_2 ⟶

b. 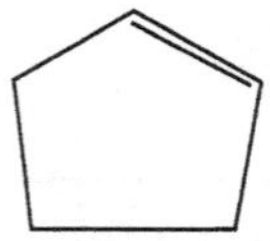   + Cl_2 ⟶

E. Identification of Unknown

Results of tests with unknown

Unknown	Combustion	Bromine Test	KMnO$_4$ Test	Alkane or Alkene?

Explain your conclusion.

Alcohols and Phenols

Goals

- Determine chemical and physical properties of alcohols and phenols.
- Classify an alcohol as primary, secondary, or tertiary.
- Perform a chemical test to distinguish between the classes of alcohols.
- Write the formulas of the oxidation products of alcohols.

Discussion

A. Structures of Alcohols and Phenol

Alcohols are organic compounds that contain the hydroxyl group (–OH). The simplest alcohol is methanol. Ethanol is found in alcoholic beverages and preservatives, and is used as a solvent. 2-Propanol, also known as rubbing alcohol, is found in astringents and perfumes.

CH_3OH

Methanol
(methyl alcohol)

$\overset{\displaystyle OH}{\underset{\displaystyle |}{CH_3CHCH_3}}$

2-Propanol
(isopropyl alcohol)

CH_3CH_2OH

Ethanol
(ethyl alcohol)

A benzene ring with a hydroxyl group is known as phenol. Concentrated solutions of phenol are caustic and cause burns. However, derivatives of phenol, such as thymol, are used as antiseptics and are sometimes found in cough drops.

Phenol

Thymol
(2-isopropyl-5-methylphenol)

Classification of Alcohols

In a primary (1°) alcohol, the carbon atom attached to the –OH group is bonded to one other carbon atom. In a secondary (2°) alcohol, it is attached to two carbon atoms and in a tertiary (3°) alcohol to three carbon atoms.

$CH_3-\overset{\displaystyle H}{\underset{\displaystyle H}{\overset{\displaystyle |}{\underset{\displaystyle |}{C}}}}-OH$

Ethanol
primary (1°) alcohol

$CH_3-\overset{\displaystyle CH_3}{\underset{\displaystyle H}{\overset{\displaystyle |}{\underset{\displaystyle |}{C}}}}-OH$

2-Propanol
secondary (2°) alcohol

$CH_3-\overset{\displaystyle CH_3}{\underset{\displaystyle CH_3}{\overset{\displaystyle |}{\underset{\displaystyle |}{C}}}}-OH$

2-Methyl-2-Propanol
tertiary (3°) alcohol

B. Properties of Alcohols and Phenol

The polarity of the hydroxyl group (–OH) makes alcohols with four or fewer carbon atoms soluble in water because they can form hydrogen bonds. However, in longer-chain alcohols, a large hydrocarbon section makes them insoluble in water.

Acidity of Phenol

In water, phenol acts as a weak acid because the hydroxyl group ionizes slightly. Although phenol has six carbon atoms, its acid behavior makes it soluble in water.

C. Oxidation of Alcohols

Primary and secondary alcohols are easily oxidized. An oxidation consists of removing an H from the –OH group and another H from the C atom attached to the –OH group. Tertiary alcohols do not undergo oxidation because there are no H atoms on that C atom. Primary and secondary alcohols can be distinguished from tertiary alcohols using a solution with chromate, CrO_4^{2-}. An oxidation has occurred when the orange color of the chromate solution turns green.

D. Ferric Chloride Test

Phenols react with the Fe^{3+} ion in a ferric chloride ($FeCl_3$) solution to give complex ions with strong colors from red to purple.

$$\text{Phenol} \quad + \quad Fe^{3+} \quad \rightarrow \quad Fe^{3+} \bullet \text{ phenol complex}$$
$$\textit{colorless} \qquad\quad \textit{yellow} \qquad\qquad\qquad\qquad\quad \textit{purple}$$

E. Identification of Unknown

The group of tests for alcohols and phenols described in this experiment will be used to identify the functional group and family of an unknown substance.

Lab Information

Time: 2 hr
Comments: Be careful when you work with chromate solution. It contains concentrated acid.
                      Do not use burners in lab when you work with flammable organic compounds.
                      Tear out the Lab report sheets and place them beside the matching procedures.
Related topics: Alcohols, classification of alcohols, solubility of alcohols in water, phenols, oxidation of alcohols

Experimental Procedures

GOGGLES MUST BE WORN!

A. Structures of Alcohols and Phenol

Materials: Organic model kits

Observe the models or obtain an organic model kit and construct models of ethanol, 2-propanol, and *t*-butyl alcohol (2-methyl-2-propanol). Write the condensed structural formula of each. Write the condensed structural formula for phenol. Classify each alcohol as a primary, secondary, or tertiary alcohol.

B.   Properties of Alcohols and Phenol

Materials: 6 test tubes, pH paper, stirring rod, ethanol, 2-propanol, t-butyl alcohol (2-methyl-2-propanol), cyclohexanol, 20% phenol, and unknown

Odor Place 5 drops of each of the alcohols, phenol, and unknown to six separate test tubes. *Avoid skin contact with phenol.* Carefully detect the odor of each. Hold your breath as you gently fan some fumes from the top of the test tube toward you.

Solubility in water Add about 2 mL of water (40 drops) to each test tube. Shake and determine whether each alcohol is soluble or not. If the substance is soluble in water, you will see a clear solution with no separate layers. If it is insoluble, a cloudy mixture or separate layer will form. Record your observations.

Acidity Obtain a container of pH paper. Place a stirring rod in one of the alcohols and touch a drop to the pH paper. Compare the color of the paper with the chart on the container to determine the pH of the solution. Record.

DISPOSE OF ORGANIC SUBSTANCES IN DESIGNATED WASTE CONTAINERS!

C. Oxidation of Alcohols

Materials: 6 test tubes, ethanol, 2-propanol, *t*-butyl alcohol (2-methyl-2-propanol), cyclohexanol, phenol, unknown, 2% chromate solution

C.1 Place 8 drops of the alcohols and an unknown in separate test tubes. Carefully add 2 drops of chromate solution to each. Look for a color change in the chromate solution as you add it to the sample. If the orange color turns to green in 1–2 minutes, oxidation of the alcohol has taken place. If the color remains orange, no reaction has occurred. If a test tube becomes hot, place it in a beaker of ice-cold water. Record your observations. Caution: **Chromate solution contains concentrated H_2SO_4, which is corrosive.**

C.2 Draw the condensed structural formula of each alcohol.

C.3 Classify each alcohol as primary (1°), secondary (2°), or tertiary (3°).

C.4 Draw the condensed structural formulas of the products where oxidation occurred. When there is no change in color, no oxidation took place. Write "no reaction" (NR).

D. Ferric Chloride Test

Materials: 6 test tubes, ethanol, 2-propanol, *t*-butyl alcohol (2-methyl-2-propanol), cyclohexanol, 20% phenol, unknown, 1% $FeCl_3$ solution

Place 5 drops of the alcohols and unknown in separate test tubes. Add 5 drops of 1% $FeCl_3$ solution to each. Stir and record observations.

DISPOSE OF ORGANIC SUBSTANCES IN DESIGNATED WASTE CONTAINERS!

E. Identification of Unknown Substance

Use the test results to identify your unknown as one of the five compounds used in this experiment.

Report Sheet - Lab 24

Date _____   Name _____

Section _____   Team _____

Instructor _____

Pre-Lab Study Questions

1. What is the functional group of an alcohol and a phenol?

2. Why are some alcohols soluble in water?

3. How are alcohols classified?

A.   Structures of Alcohols and Phenols

Ethanol	2-Propanol
Classification:	
t-Butyl alcohol (2-methyl-2-propanol)	Phenol
Classification:	

Report Sheet - Lab 24

Questions and Problems

Q.1 Write the structures and classifications of the following alcohols:

1-Pentanol	3-Pentanol
Cyclopentanol	1-Methylcyclopentanol

Q.2 In your textbook or a chemistry handbook, look up and draw the condensed structural formulas and uses of thymol, menthol, and resorcinol. Circle the phenol functional group in each structure. If the compounds are available in lab, carefully note and describe their odors.

B. Properties of Alcohols and Phenols

Alcohol	Odor	Soluble in Water?	pH
Ethanol			
2-propanol			
t-butyl alcohol			
Cyclohexanol			
Phenol			
Unknown			

Report Sheet - Lab 24

C.   Oxidation of Alcohols

Alcohol	C.1 Color Change with CrO_4^{2-}	C.2 Condensed Structural Formula	C.3 Classification	C.4 Oxidation Product (If reaction takes place)
Ethanol				
2-Propanol				
t-butyl alcohol				
Cyclohexanol				
Phenol			╳	
Unknown		╳	╳	╳

Questions and Problems

Q.3   Write the product of the following reactions (if no reaction, write NR):

a. $CH_3CH_2CH_2OH$ $\xrightarrow{[O]}$

b. $CH_3\overset{\displaystyle OH}{\underset{|}{C}H}CH_2CH_3$ $\xrightarrow{[O]}$

c.   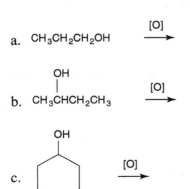 $\xrightarrow{[O]}$

Report Sheet - Lab 24

D. Ferric Chloride Test

Alcohol	FeCl₃ Test
Ethanol	
2-propanol	
t-Butyl alcohol	
Cyclohexanol	
Phenol	
Unknown	

E. Identification of Unknown Substance

Unknown # _____

Summary of Testing	Results	Conclusions
B. Odor		
B. Soluble in water?		
B. pH		
C. Oxidation: CrO_4^{2-}		
D. $FeCl_3$		
Name of Unknown	**Structure**	

Aldehydes and Ketones

Goals

- Write the functional groups of aldehydes and ketones.
- Determine chemical and physical properties of aldehydes and ketones.
- Perform chemical tests to distinguish between aldehydes and ketones.

Discussion

A. Structures of Some Aldehydes and Ketones

Aldehydes and ketones both contain the carbonyl group. In an aldehyde, the carbonyl group has a hydrogen atom attached; the aldehyde functional group occurs at the end of the carbon chain. In a ketone, the carbonyl group is located between two of the carbon atoms within the chain.

Aldehydes

Ketone

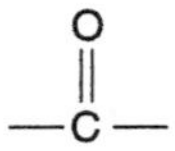

Carbonyl functional group

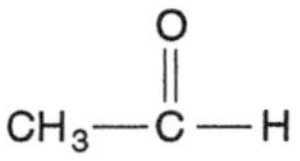

Acetaldehyde (ethanal)

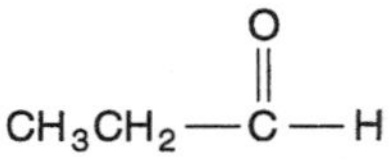

Propionaldehyde (propanal)

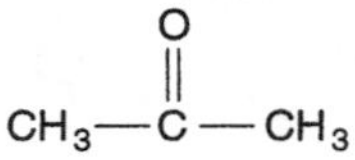

Acetone (2-propanone)

B. Properties of Aldehydes and Ketones

Many aldehydes and ketones have sharp odors. If you have taken a biology class, you may have noticed the odor of Formalin™, which is a solution of formaldehyde. When you remove fingernail polish, you may notice the strong odor of acetone, the simplest ketone, which is used as the solvent. Aromatic aldehydes have a variety of odors. Benzaldehyde, the simplest aromatic aldehyde, has an odor of almonds.

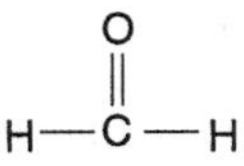

Formaldehyde

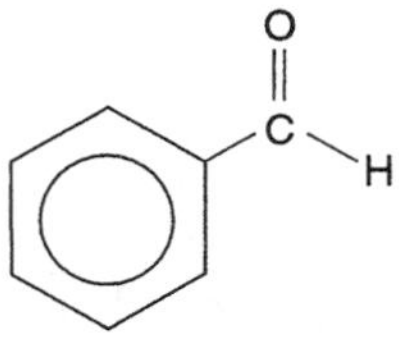

Benzaldehyde

C. Iodoform Test for Methyl Ketones

Ketones containing a methyl group attached to the carbonyl give a reaction with iodine (I_2) in a NaOH solution. The reaction produces solid, yellow iodoform, CHI_3. Iodoform, which has a strong medicinal odor, is used as an antiseptic.

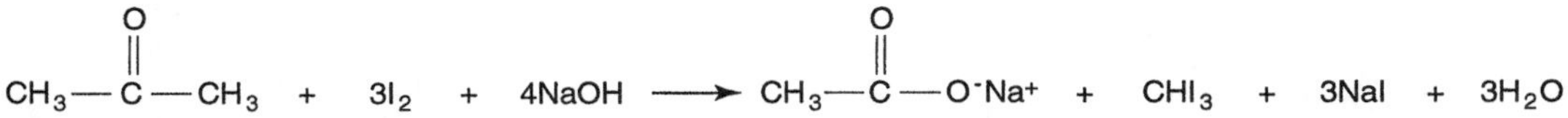

$$CH_3-\overset{\overset{O}{\|}}{C}-CH_3 + 3I_2 + 4NaOH \longrightarrow CH_3-\overset{\overset{O}{\|}}{C}-O^-Na^+ + CHI_3 + 3NaI + 3H_2O$$

Methyl ketone   Iodine
(red)

Iodoform
(yellow)

D. Oxidation of Aldehydes and Ketones

Aldehydes are oxidized using Benedict's solution, which contains cupric ion, Cu^{2+}. Because ketones cannot oxidize, this test can distinguish aldehydes from ketones. In the oxidation reaction, the blue-green Cu^{2+} is reduced to cuprous ion (Cu^+), which forms a reddish-orange precipitate of Cu_2O.

$$CH_3-\overset{\overset{\displaystyle O}{\|}}{C}-H \;+\; 2Cu^{2+} \longrightarrow CH_3-\overset{\overset{\displaystyle O}{\|}}{C}-OH \;+\; Cu_2O(s)$$

Aldehyde   Blue   Red-orange

$$CH_3-\overset{\overset{\displaystyle O}{\|}}{C}-CH_3 \;+\; 2Cu^{2+} \longrightarrow \text{No reaction (stays blue)}$$

Ketone   Blue

E. Identification of an Unknown

Using the results of the tests, an unknown substance can be identified as an aldehyde or ketone.

Lab Information

Time: 2 hr

Comments: Flammable compounds are used in this experiment. Do not use burners.
In tests with color changes, carefully observe the color of the reactants before they are mixed.
Tear out the Lab report sheets and place them beside the matching procedures.

Related topics: Aldehydes, ketones, oxidation of aldehydes

Experimental Procedures

BE SURE TO WEAR YOUR GOGGLES!

A. Structures of Some Aldehydes and Ketones

Materials: Organic model kits

Make models or observe models of formaldehyde, acetaldehyde, propionaldehyde, acetone, butanone, and cyclohexanone. Draw their condensed structural formulas. Write the IUPAC and common names (if any) for each.

B. Properties of Aldehydes and Ketones

Materials: Chemistry handbook, test tubes, droppers, 5- or 10-mL graduated cylinder, acetone, benzaldehyde, camphor, vanillin, cinnamaldehyde, 2,3-butanedione, propionaldehyde, cyclohexanone, and unknown

Odors of Aldehydes and Ketones

B.1 Carefully detect the odor of samples of acetone, benzaldehyde, camphor, vanillin, cinnamaldehyde, and 2,3-butanedione, and unknown.

B.2 Draw their condensed structural formulas. You may need a chemistry handbook or a *Merck Index*. Identify each as a ketone or aldehyde.

Solubility of Aldehydes and Ketones

B.3 Place 2 mL of water in each of 4 separate test tubes. Add 5 drops of propionaldehyde (propanal), benzaldehyde, acetone, cyclohexanone, and unknown. Record your observations. *Save the samples for part C.*

C. Iodoform Test for Methyl Ketones

Materials: Test tubes from part B.3, dropper, 10% NaOH, warm water bath, and iodine test reagent

Using the test tubes from part B.3, add 10 drops of 10% NaOH to each. Warm the tubes in a warm water bath to 50–60°C. Add 20 drops of iodine test reagent. Look for the formation of a yellow solid precipitate. Record your results.

D. Oxidation of Aldehydes and Ketones

Materials: Test tubes, propionaldehyde (propanal), benzaldehyde, acetone, cyclohexanone, unknown, benedict's reagent, droppers, boiling water bath

Place 10 drops of propionaldehyde (propanal), benzaldehyde, acetone, cyclohexanone and unknown in separate test tubes. Label. Add 2 mL of Benedict's reagent to each test tube. Place the test tubes in a boiling water bath for 5 minutes. The appearance of the red-orange color of Cu_2O indicates that oxidation has occurred. Moderate amounts of Cu_2O will blend with the blue Cu^{2+} solution to form green or rust color. Record your observations. Identify the compounds that gave an oxidation reaction.

E. Identification of an Unknown

If you were given an unknown compound, you can now compare the results of the tests for the unknown with the tests you performed with known aldehydes and ketones. Identify your unknown as an aldehyde or a ketone.

Report Sheet - Lab 25

Date _____   Name _____

Section _____   Team _____

Instructor _____

Pre-Lab Study Questions

1. What is the functional group of an aldehyde? A ketone?

2. What is the oxidation product of an aldehyde?

A.   Structures of Some Aldehydes and Ketones

Formaldehyde IUPAC Name _____	Acetaldehyde IUPAC Name_____
Propionaldehyde IUPAC Name_____	Acetone IUPAC Name_____
Butanone Common Name_____	Cyclohexanone

Report Sheet - Lab 25

B. Properties of Aldehydes and Ketones

	B.1 Odor	B.2 Condensed Structural Formula	Aldehyde or Ketone?
Acetone			
Benzaldehyde			
Camphor			
Vanillin			
Cinnamaldehyde			
2,3-butanedione			
Unknown			

Report Sheet - Lab 25

Questions and Problems

Q.1 What aldehyde or ketone might be present in the following everyday products?

Artificial butter flavor in popcorn _____

Almond-flavored cookies _____

Candies with cinnamon flavor _____

Nail polish remover _____

B., C., and D. Solubility, Iodoform, and Oxidation of Aldehydes and Ketones

	B.3 **Solubility** Soluble in water?	C. **Iodoform Test** Methyl ketone present?	D. **Benedict's Test** Oxidation occurred?
Propionaldehyde			
Benzaldehyde			
Acetone			
Cyclohexanone			
Unknown			

Questions and Problems

Q.2 Complete the following with the word *soluble or insoluble:*

Aldehydes and ketones containing one to four carbon atoms are _____ in water.

Aldehydes and ketones containing five or more carbon atoms are _____ in water.

Report Sheet - Lab 25

Q.3 Indicate the test results for each of the following compounds in the iodoform test and in the Benedict's test:

	Iodoform Test	Benedict's Test
$CH_3CCH_2CH_3$ (O)		
CH_3CH (O)		
$CH_3CH_2CCH_2CH_3$ (O)		
CH_3CCH_2CH (O O)		

Q.4 Two compounds, A and B, have the formula of C_3H_6O. Determine their condensed structural formulas and names using the following test results.

a. Compound A forms a red-orange precipitate with Benedict's reagent but does not react with iodoform.

b. Compound B forms a yellow solid in the iodoform test but does not react with Benedict's reagent.

Report Sheet - Lab 25

E.   Identification of an Unknown Substance

Unknown _____

Summary of Testing	Results	Conclusion
Odor		Does it have a familiar odor?
Solubility in water		How many carbon atoms?
Iodoform test		Is it a methyl ketone?
Benedict's test		Is it an aldehyde?

Questions and Problems

Q.5   What can you conclude about the structure and functional group of your unknown? Explain.

Q.6   What chemical tests could you use to distinguish between 2-pentanone and 3-pentanone?

Goals

- Identify the characteristic functional groups of carbohydrates.
- Describe common carbohydrates and their sources.
- Distinguish between monosaccharides, disaccharides, and polysaccharides.

Discussion

Carbohydrates in our diet are our major source of energy. Foods high in carbohydrates include potatoes, bread, pasta, and rice. If we take in more carbohydrate than we need for energy, the excess is converted to fat, which can lead to a weight gain. The carbohydrate family is organized into three classes, which are the monosaccharides, disaccharides, and polysaccharides.

A.   Monosaccharides

Monosaccharides contain C, H, and O in units of $(CH_2O)_n$. Most common monosaccharides have six carbon atoms (hexoses) with a general formula of $C_6H_{12}O_6$. They contain many hydroxyl groups (–OH) along with a carbonyl group. The aldoses are monosaccharides with an aldehyde group, and ketoses contain a ketone group.

Monosaccharides		Sources
Glucose	$C_6H_{12}O_6$	Fruit juices, honey, corn syrup
Galactose	$C_6H_{12}O_6$	Lactose hydrolysis
Fructose	$C_6H_{12}O_6$	Fruit juices, honey, sucrose hydrolysis

Glucose, a hexose, is the most common monosaccharide; it is also known as blood sugar.

D-Glucose

The letter D refers to the orientation of the hydroxyl group on the chiral carbon that is farthest from the carbonyl group at the top of the chain (carbon 1). The D- and L-isomers of glyceraldehyde illustrate the position of the –OH on the central, chiral atom.

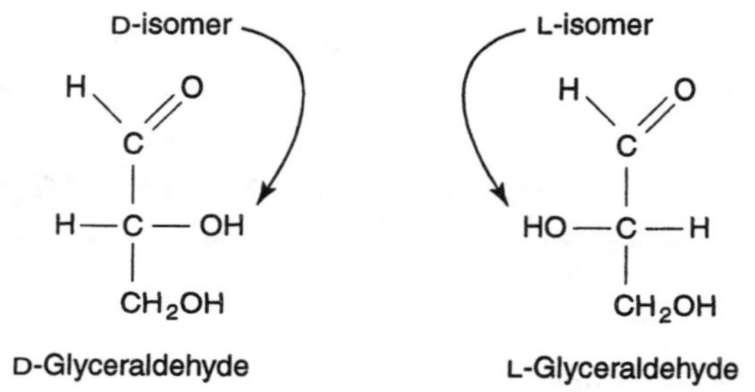

D-Glyceraldehyde   L-Glyceraldehyde

Haworth Structures

Most of the time glucose exists in a ring structure, which forms when the OH on carbon 5 forms a hemiacetal bond with the aldehyde group. In the Haworth structure the new hydroxyl group on carbon 1 may be drawn above carbon 1 (the β form) or below carbon 1 (the α form).

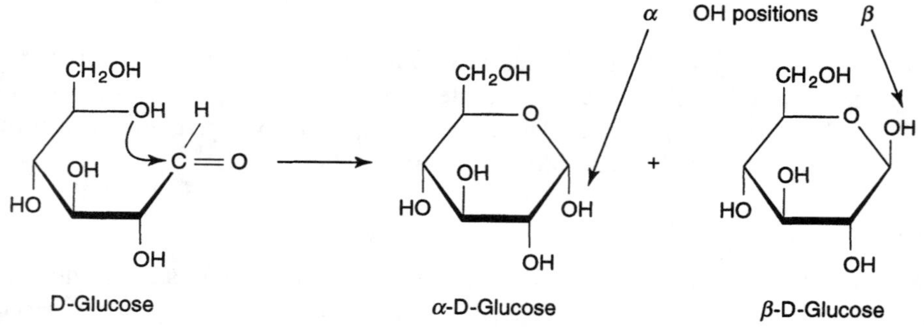

D-Glucose α-D-Glucose β-D-Glucose

B. Disaccharides

The disaccharides contain two of the common monosaccharides. Some common disaccharides include maltose, sucrose (table sugar), and lactose (milk sugar).

Disaccharides	Sources	Monosaccharides
Maltose	Germinating grains, starch hydrolysis	Glucose + glucose
Lactose	Milk, yogurt, ice cream	Glucose + galactose
Sucrose	Sugar cane, sugar beets	Glucose + fructose

In a disaccharide, two monosaccharides form a glycosidic bond with the loss of water. For example, in maltose, two glucose units are linked by an α-1,4-glycosidic bond.

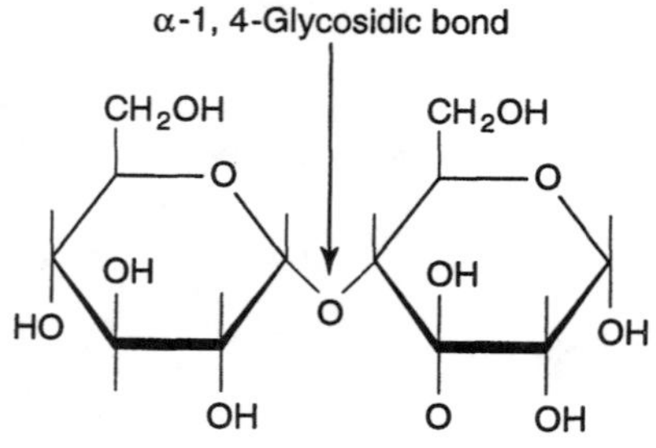

α-Maltose

C. Polysaccharides

Polysaccharides are long-chain polymers that contain many thousands of monosaccharides (usually glucose units) joined together by glycosidic bonds. Three important polysaccharides are starch, cellulose, and glycogen. They all contain glucose units, but differ in the type of glycosidic bonds and the amount of branching in the molecule.

Polysaccharides	Found in	Monosaccharides
Starch (amylose, amylopectin)	Rice, wheat, grains, cereals	Glucose
Glycogen	Muscle, liver	Glucose
Cellulose	Wood, plants, paper, cotton	Glucose

Starch is an insoluble storage form of glucose found in rice, wheat, potatoes, beans, and cereals. Starch is composed of two kinds of polysaccharides, amylose and amylopectin. *Amylose*, which makes up about 20% of starch, consists of α-D-glucose molecules connected by α-1,4-glycosidic bonds in a continuous chain. A typical polymer of amylose may contain from 250 to 4000 glucose units.

α-1,4-Glycosidic bonds in amylose

Amylopectin is a branched-chain polysaccharide that makes up as much as 80% of starch. In amylopectin, α-1,4-glycosidic bonds connect most of the glucose molecules. However, at about every 25 glucose units, there are branches of glucose molecules attached by α-1,6-glycosidic bonds between carbon 1 of the branch and carbon 6 in the main chain.

Amylopectin

Glycogen is a similar to amylopectin but it is even more highly branched, with α-1,6-glycosidic bonds about every 10-15 glucose units.

Cellulose is the major structural material of wood and plants. Cotton is almost pure cellulose. In cellulose, glucose molecules form a long unbranched chain similar to amylose except that β-1,4-glycosidic bonds connect the glucose molecules. The β isomers are aligned in parallel rows that are held in place by hydrogen bonds between the rows. This gives a rigid structure for cell walls in wood and fiber and makes cellulose more resistant to hydrolysis.

Cellulose

Lab Information

Time: 2 hr

Comments: Tear out the report sheets and place them next to the matching procedures. In the study of carbohydrates, it is helpful to review stereoisomers and the formation of hemiacetals.

Related Topics: Carbohydrates, monosaccharides, disaccharides, polysaccharides, hemiacetals, stereoisomers, aldohexoses, ketohexoses, chiral compounds, Fischer projection, Haworth structures

Experimental Procedures

A. Monosaccharides

Materials: Organic model kits or prepared models

A.1   Make or observe models of L-glyceraldehyde and D-glyceraldehyde. Draw the Fischer projections.

A.2   Draw the Fischer projection for D-glucose. Draw the Haworth (cyclic) formulas for the α and β anomers.

A.3   Draw the Fischer projections for D-fructose and D-galactose. Draw the Haworth (cyclic) formulas for the α anomers of each.

B. Disaccharides

B.1   Using Haworth formulas, write the structure for α-D-maltose. Look at a model if available.

B.2   Write an equation for the hydrolysis of α-D-maltose by adding H_2O to the glycosidic bond.

B.3   Using Haworth formulas, write an equation for the formation of α-D-lactose from β-D-galactose and α-D-glucose.

B.4   Draw the structure of sucrose and circle the glycosidic bond.

C. Polysaccharides

C.1   Draw a portion of amylose using four units of α-D-glucose. Indicate the glycosidic bonds.

C.2   Describe how the structure of amylopectin differs from the structure of amylose.

C.3   Draw a portion of cellulose using four units of β-D-glucose. Indicate the glycosidic bonds.

Report Sheet - Lab 26

Date _____   Name _____

Section _____   Team _____

Instructor _____

Pre-Lab Study Questions

1. What are some sources of carbohydrates in your diet?

2. What does the D in D-glucose mean?

3. What is the bond that links monosaccharides in di- and polysaccharides?

A.   Monosaccharides

A.1   Fischer projections

L-glyceraldehyde   D-glyceraldehyde

How does L-glyceraldehyde differ from D-glyceraldehyde?

A.2   Fischer projection of D-glucose   Haworth (cyclic) formulas

α-D-glucose   β-D-glucose

Report Sheet - Lab 26

A.3 Fischer projection of D-fructose

Haworth (cyclic) formula for α-D-fructose

Fischer projection of D-galactose

Haworth (cyclic) formula for α-D-galactose

Questions and Problems

Q.1 How does the structure of D-glucose compare to the structure of D-galactose?

B. Disaccharides

B.1 Structure of α-D-maltose

B.2 Equation for the hydrolysis of α-D-maltose

Report Sheet - Lab 26

B.3 Formation of α-D-lactose

B.4 Structure of sucrose

Questions and Problems

Q.2 What is the type of glycosidic bond in maltose?

Q.3 Why does maltose have both α and β anomers? Explain.

C. Polysaccharides

C.1 A portion of amylose

C.2 Comparison of amylopectin to amylose

Report Sheet - Lab 26

C.3   A portion of cellulose

Questions and Problems

Q.4   What is the monosaccharide that results from the complete hydrolysis of amylose?

Q.5   What is the difference in the structure of amylose and cellulose?

Tests for Carbohydrates

Goals

- Observe physical and chemical properties of some common carbohydrates.
- Use physical and chemical tests to distinguish between monosaccharides, disaccharides, and polysaccharides.
- Identify an unknown carbohydrate.
- Relate the process of digestion to the hydrolysis of carbohydrates.

Discussion

A. Benedict's Test for Reducing Sugars

All of the monosaccharides and most of the disaccharides can be oxidized. When the cyclic structure opens, the aldehyde group is available for oxidation. Benedict's reagent contains Cu^{2+} ion that is reduced. Therefore, all the sugars that react with Benedict's reagent are called *reducing sugars*. Ketoses also act as reducing sugars because the ketone group on carbon 2 isomerizes to give an aldehyde group on carbon 1.

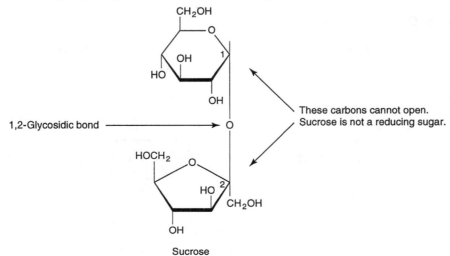

When oxidation of a sugar occurs, the Cu^{2+} is reduced to Cu^{+}, which forms a red precipitate of cuprous oxide, $Cu_2O(s)$. The color of the precipitate varies from green to gold to red depending on the concentration of the reducing sugar.

$$\underset{\text{Reducing sugar}}{\text{sugar}-\overset{\overset{\displaystyle O}{\|}}{C}H} \quad + \quad \underset{\text{Blue}}{2Cu^{2+}} \quad \overset{OH^{-}}{\longrightarrow} \quad \underset{\text{Oxidized sugar}}{\text{sugar}-\overset{\overset{\displaystyle O}{\|}}{C}OH} \quad + \quad \underset{\text{Blue}}{Cu_2O(s)}$$

Sucrose is not a reducing sugar because it cannot revert to the open-chain form that would provide the aldehyde group needed to reduce the cupric ion.

B. Seliwanoff's Test for Ketoses

Seliwanoff's test is used to distinguish between hexoses with a ketone group and hexoses that are aldehydes. With ketoses, a deep red color is formed rapidly. Aldoses give a light pink color that takes a longer time to develop. The test is most sensitive for fructose, which is a ketose.

C. Fermentation Test

Most monosaccharides and disaccharides undergo fermentation in the presence of yeast. The products of fermentation are ethyl alcohol (CH_3CH_2OH) and carbon dioxide (CO_2). The formation of bubbles of carbon dioxide is used to confirm the fermentation process.

$$C_6H_{12}O_6 \xrightarrow{\text{yeast}} 2C_2H_5OH + 2CO_2(g)$$
$$\text{Glucose} \qquad\qquad \text{Ethanol}$$

Although enzymes are present for the hydrolysis of most disaccharides, they are not available for lactose. The enzymes needed for the fermentation of galactose are not present in yeast. Lactose and galactose give negative results with the fermentation test.

D. Iodine Test for Polysaccharides

When iodine (I_2) is added to amylose, the helical shape of the unbranched polysaccharide traps iodine molecules, producing a deep blue-black complex. Amylopectin, cellulose, and glycogen react with iodine to give red to brown colors. Glycogen produces a reddish-purple color. Monosaccharides and disaccharides are too small to trap iodine molecules and do not form dark colors with iodine.

E. Hydrolysis of Disaccharides and Polysaccharides

Disaccharides hydrolyze in the presence of an acid to give the individual monosaccharides.

$$\text{Sucrose} + H_2O \xrightarrow{H^+} \text{Glucose} + \text{Fructose}$$

In the laboratory, we use water and acid to hydrolyze starches, which produce smaller saccharides such as maltose. Eventually, the hydrolysis reaction converts maltose to glucose molecules. In the body, enzymes in our saliva and from the pancreas carry out the hydrolysis. Complete hydrolysis produces glucose, which provides about 50% of our nutritional calories.

$$\text{Amylose, amylopectin} \xrightarrow[\text{amylase}]{H^+ \text{ or}} \text{dextrins} \xrightarrow[\text{amylase}]{H^+ \text{ or}} \text{maltose} \xrightarrow[\text{maltase}]{H^+ \text{ or}} \text{many D-glucose units}$$

F. Testing Foods for Carbohydrates

Several of the tests such as the iodine test can be carried out with food products such as cereals, bread, crackers, and pasta. Some of the carbohydrates we have discussed can be identified.

Lab Information

Time: 3 hr
Comments: Tear out the report sheets and place them next to the matching procedures. .
Related Topics: Carbohydrates, hemiacetals, aldohexoses, ketohexoses, reducing sugars, fermentation

Experimental Procedures

A. Benedict's Test for Reducing Sugars

Materials: Test tubes, 400-mL beaker, droppers, hot plate or Bunsen burner, 5- or 10-mL graduated cylinder, Benedict's reagent, 2% carbohydrate solutions: glucose, fructose, sucrose, lactose, starch, and an unknown

Place 10 drops of solutions of glucose, fructose, sucrose, lactose, starch, water, and unknown in separate test tubes. Label each test tube. Add 2 mL of Benedict's reagent to each sample. Place the test tubes in a boiling water bath for 3–4 minutes. The formation of a greenish to reddish-orange color indicates the presence of a reducing sugar. If the solution is the same color as the Benedict's reagent in water (the control), there has been no oxidation reaction. Record your observations. Classify each as a reducing or nonreducing sugar.

B. Seliwanoff's Test for Ketoses

Materials: Test tubes, 400-mL beaker, droppers, hot plate or Bunsen burner, 5- or 10-mL graduated cylinder, Seliwanoff's reagent, 2% carbohydrate solutions: glucose, fructose, sucrose, lactose, starch, and an unknown

Place 10 drops of solutions of glucose, fructose, sucrose, lactose, starch, water, and unknown in separate test tubes. Add 2 mL of Seliwanoff's reagent to each. ***The reagent contains concentrated HCl. Use carefully.***

Place the test tubes in a boiling hot water bath and note the time. After 1 minute, observe the colors in the test tubes. A rapid formation of a deep red color indicates the presence of a ketose. Record your results as a fast color change, slow change, or no change.

C. Fermentation Test

Materials: Fermentation tubes (or small and large test tubes), baker's yeast, 2% carbohydrate solutions: glucose, fructose, sucrose, lactose, starch, and an unknown

Fill fermentation tubes with a solution of glucose, fructose, sucrose, lactose, starch, water, and unknown. Add 0.2 g of yeast to each and mix well. See Figure 27.1.

Figure 27.1 Fermentation tube filled with a carbohydrate solution

If fermentation tubes are not available, use small test tubes placed upside down in larger test tubes. Cover the mouth of the large test tube with filter paper or cardboard. Place your hand firmly over the paper cover and invert. When the small test tube inside has completely filled with the mixture, return the larger test tube to an upright position. See Figure 27.2.

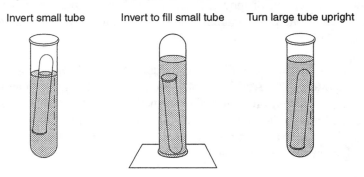

Invert small tube   Invert to fill small tube   Turn large tube upright

Figure 27.2   Test tubes used as fermentation tubes

Set the tubes aside. At the end of the laboratory period, and again at the next laboratory period, look for gas bubbles in the fermentation tubes or inside the small tubes. Record your observations. See Figure 27.3.

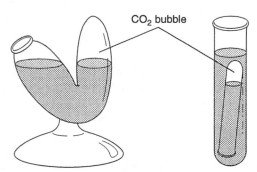

CO_2 bubble

Figure 27.3   Fermentation tubes with CO_2 bubbles

D.   Iodine Test for Polysaccharides

Materials: Spot plate or test tubes, droppers, iodine reagent, 2% carbohydrate solutions in dropper bottles: glucose, fructose, sucrose, lactose, starch, and an unknown

Using a spot plate, place 5 drops of each solution of glucose, fructose, sucrose, lactose, starch, water, and unknown in the wells. (If you do not have a spot plate, use small test tubes.) Add 1 drop of iodine solution to each sample. A dark blue-black color is a positive test for amylose in starch. A red or brown color indicates the presence of other polysaccharides. Record your results. Complete the table to identify your unknown.

E.   Hydrolysis of Disaccharides and Polysaccharides

Materials: Test tubes, 10-mL graduated cylinder, 400-mL beaker (boiling water bath), hot plate or Bunsen burner, spot plate or watch glass, 10% HCl, 10% NaOH, red litmus paper, iodine reagent, Benedict's reagent, 2% starch and sucrose solutions in dropper bottles

Place 3 mL of 2% starch in two test tubes and 3 mL of 2% sucrose solution in two more test tubes. To one sample each of sucrose and starch, add 20 drops of 10% HCl. To the other samples of sucrose and starch, add 20 drops of H_2O. Label the test tubes and heat in a boiling water bath for 10 minutes.

Remove the test tubes from the water bath and let them cool. To the samples containing HCl, add 10% NaOH (about 20 drops) until one drop of the mixture turns litmus paper blue, indicating the HCl has been neutralized. Test the samples for hydrolysis as follows:

Iodine Test Place 5 drops of each solution on a spot plate or watch glass. Add 1 drop of iodine reagent to each. Record observations. Determine if hydrolysis has occurred in each.

Benedict's Test Add 2 mL of Benedict's reagent to each of the samples and heat in a boiling water bath for 3–4 minutes. Determine if hydrolysis has occurred in each.

F. Testing Foods for Carbohydrates

Materials: Sugar samples (refined, brown, "natural," powdered), honey, syrups (corn, maple, fruit), foods with starches: cereals, pasta, bread, crackers, potato, Benedict's solution, Seliwanoff's reagent, iodine reagent

Obtain two carbohydrate samples to test. Perform the Benedict's, Seliwanoff's, and iodine tests on each. Describe the kinds of carbohydrates you identify in each sample.

Carboxylic Acids and Esters

Goals

- Write the structural formulas of carboxylic acids and esters.
- Determine the solubility and acidity of carboxylic acids and their salts.
- Write equations for neutralization and esterification of acids.
- Prepare esters and identify their characteristic odors.

Discussion

A. Carboxylic Acids and Their Salts

A salad dressing made of oil and vinegar tastes tart because it contains vinegar, which is known as acetic acid (ethanoic acid). The sour taste of fruits such as lemons is due to acids such as citric acid. Face creams contain alpha hydroxy acids such as glycolic acid. All these acids are carboxylic acids, which contain the carboxyl group: a carbonyl group attached to a hydroxyl group. A dicarboxylic acid such as malonic acid, found in apples, has two carboxylic acid functional groups. The carboxylic acid of benzene is called benzoic acid.

$$
\begin{array}{c}
\overset{\displaystyle O}{\overset{\|}{-C}} -OH \\
\text{Carboxyl group}
\end{array}
$$

$$
\begin{array}{cccc}
\overset{\displaystyle O}{\underset{\text{Acetic acid}}{CH_3\overset{\|}{C}-OH}} &
\overset{\displaystyle O}{\underset{\text{Propionic acid}}{CH_3CH_2\overset{\|}{C}-OH}} &
\overset{\displaystyle O\qquad O}{\underset{\text{Malonic acid}}{HO-\overset{\|}{C}-CH_2-\overset{\|}{C}-OH}} &
\underset{\text{Benzoic acid}}{\overset{\displaystyle O}{\overset{\|}{C}-OH}}
\end{array}
$$

Ionization of Carboxylic Acids in Water

Carboxylic acids are weak acids because the carboxylic acid group ionizes slightly in water to give a proton and a carboxylate ion. However, like the alcohols, the polarity of the carboxylic acid group makes acids with one to four carbon atoms soluble in water. Acids with two or more carboxyl groups (diacids) are more soluble in water.

$$
\underset{\text{Acetic acid}}{CH_3-\overset{\displaystyle O}{\overset{\|}{C}}-OH} \; + \; H_2O \;\rightleftharpoons\; \underset{\substack{\text{Acetate ion}\\\text{(a carboxylate ion)}}}{CH_3-\overset{\displaystyle O}{\overset{\|}{C}}-O^-} \; + \; H_3O^+
$$

Neutralization of Carboxylic Acids

An important feature of carboxylic acids is their neutralization by bases such as sodium hydroxide to form carboxylate salts and water. We saw in an earlier experiment that neutralization is the reaction of an acid with a base to give a salt and water. Even insoluble carboxylic acids with five or more carbon atoms can be neutralized to give corresponding salts that are usually soluble in water. For this reason, acids used in food products or medications are in their soluble salt form rather than the acid itself.

$$HX \quad + \quad NaOH \quad \longrightarrow \quad Na^+X^- \quad + \quad H_2O$$

Acid	Base		Salt		Water

$$CH_3-\overset{\overset{\displaystyle O}{\|}}{C}-OH \quad + \quad NaOH \quad \longrightarrow \quad CH_3-\overset{\overset{\displaystyle O}{\|}}{C}-O^-Na^+ \quad + \quad H_2O$$

Acetic acid

Sodium acetate
(a carboxylate salt)

B. Esters

Carboxylic acids may have tart or unpleasant odors, but many esters have pleasant flavors and fragrant odors. Octyl acetate gives oranges their characteristic odor and flavor; pear flavor is due to pentyl acetate. The flavor and odor of raspberries come from isobutyl formate.

$$CH_3(CH_2)_7O-\overset{\overset{\displaystyle O}{\|}}{C}-CH_3 \qquad CH_3(CH_2)_4-O-\overset{\overset{\displaystyle O}{\|}}{C}-CH_3 \qquad CH_3-\overset{\overset{\displaystyle CH_3}{|}}{C}HCH_2-O-\overset{\overset{\displaystyle O}{\|}}{C}-H$$

Octyl acetate
(oranges)

Pentyl acetate
(pears)

Isobutyl formate
(raspberries)

An ester of salicylic acid is methyl salicylate, which gives the flavor and odor of oil of wintergreen used in candies and ointments for sore muscles. When salicylic acid reacts with acetic anhydride, acetylsalicylic acid (ASA) is formed, which is aspirin, widely used to reduce fever and inflammation.

Methyl salicylate
(wintergreen)

Acetylsalicylic acid
(aspirin)

Esterification and Hydrolysis

In a reaction called *esterification,* the carboxylic acid group combines with the hydroxyl group of an

Esterification ⟶

$$CH_3-\overset{\overset{\displaystyle O}{\|}}{C}-OH \quad + \quad HO-(CH_2)_4-CH_3 \xrightleftharpoons{H^+} CH_3-\overset{\overset{\displaystyle O}{\|}}{C}-O-(CH_2)_4-CH_3 \quad + \quad H_2O$$

Acetic acid

1-Pentanol

Pentyl acetate (pear flavor)

◀——— Hydrolysis

alcohol. The reaction, which takes place in the presence of an acid, produces an ester and water. The reverse reaction, hydrolysis, occurs when an acid catalyst and water cause the decomposition of an ester to yield the carboxylic acid and alcohol. The ester product is favored when an excess of acid or alcohol is used; hydrolysis is favored when more water is used.

C. Hydrolysis of Esters

When an ester is hydrolyzed in the presence of a base, the reaction is called saponification. The products are the salt of the carboxylic acid and the alcohol. Although the ester is usually insoluble in water, the salt and alcohol (if short-chain) are soluble.

Ester bond splits

$$CH_3-\overset{\overset{\displaystyle O}{\|}}{C}-O-CH_2CH_3 \ + \ NaOH \ \longrightarrow \ CH_3-\overset{\overset{\displaystyle O}{\|}}{C}-O^-Na^+ \ + \ HO-CH_2CH_3$$

Ethyl acetate (ester)   Sodium acetate (carboxylate salt)   Ethanol (alcohol)

Lab Information

Time: 2 hr
Comments: When noting odors, hold your breath and fan across the top of a test tube to detect the odor. The formation of esters requires concentrated acid. Use carefully.
                 Tear out the report sheets and place them beside the matching procedure.
Related Topics: Carboxylic acid functional group, ester functional group, ionization of carboxylic acids, neutralization, esterification, hydrolysis, saponification

Experimental Procedures

BE SURE TO WEAR YOUR SAFETY GOGGLES!

A. Carboxylic Acids and Their Salts

Materials: Test tubes, glacial acetic acid, benzoic acid(*s*), dropper, spatula, pH paper, red and blue litmus paper, stirring rod, 400-mL beaker, hot plate or Bunsen burner, 10% NaOH, 10% HCl

A.1 Write the structural formulas for acetic acid and benzoic acid.

A.2 Place about 2 mL of water in two test tubes. Add 5 drops of acetic acid to one test tube and a small amount of benzoic solid (enough to cover the tip of a spatula) to the other. Tap the sides of the test tubes to mix or stir with a stirring rod. Identify the acid that dissolves.

A.3 Test the pH of each carboxylic acid by dipping a stirring rod into the solution, then touching it to a piece of pH paper. Compare the color on the paper with the color chart on the container and report the pH.

A.4 Place the test tube of benzoic acid (solid should be present) in a hot water bath and heat for 5 minutes. Describe the effect of heating on the solubility of acid. Allow the test tube to cool. Record your observations.

A.5 Add about 10 drops of NaOH to each test tube until a drop of the solution turns red litmus paper blue. Record your observations. Write the equations for the reactions with NaOH including the structures of the sodium salts formed.

A.6 Add about 10 drops of HCl to each sample until it is neutralized (blue litmus paper turns red). Record your observations. Write equations for the reactions of the sodium salts that formed.

B. Esters

Materials: Organic model set, test tubes, hot plate or Bunsen burner, 400-mL beaker, stirring rod, spatula, small beaker, methanol, 1-pentanol, 1-octanol, benzyl alcohol, 1-propanol, salicylic acid(s), glacial acetic acid, H_3PO_4

B.1 Make a model of acetic acid and methyl alcohol. Remove the components of water and form an ester bond to give methyl acetate. Write the equation for the formation of the ester.

B.2 As assigned, prepare one of the mixtures listed by placing 3 mL of the alcohol in a test tube and label it with the mixture number. Add 2 mL of a carboxylic acid or the amount of solid that covers the tip of a spatula. Write the condensed structural formulas for the alcohols and carboxylic acids.

Mixture	Alcohol	Carboxylic Acid
1	Methanol	Salicylic acid
2	1-Pentanol	Acetic acid
3	1-Octanol	Acetic acid
4	Benzyl alcohol	Acetic acid
5	1-Propanol	Acetic acid

Caution: Use care in dispensing glacial acetic acid. It can cause burns and blisters on the skin.

B.3 With the test tube pointed away from you, *cautiously* add 15 drops of concentrated phosphoric acid, H_3PO_4. Stir. Place the test tube in a boiling water bath for 15 minutes. Remove the test tube and *cautiously* fan the vapors toward you. Record the odors you detect such as pear, banana, orange, raspberry, or oil of wintergreen (spearmint). For a stronger odor, place 15 mL of hot water in a small beaker and pour the ester solution into the hot water.

If specified by your instructor, repeat the esterification with other mixtures of alcohol and carboxylic acid. Note the odors of esters produced by other students. Write the condensed structural formulas and names of the esters produced. *Dispose of the ester products as instructed*.

C. Hydrolysis of Esters

Materials: Test tube, test tube holder, droppers, stirring rod, hot plate or Bunsen burner, 250- or 400-mL beaker, methyl salicylate in a dropper bottle, 10% NaOH, 10% HCl, blue litmus paper

C.1 Draw the condensed structural formula of methyl salicylate.

C.2 Place 3 mL of water in a test tube. Add 5 drops of methyl salicylate. Record the appearance and odor of the ester.

C.3 Add 1 mL (20 drops) of 10% NaOH. There should be two layers in the test tube. Place the test tube in a boiling water bath for 30 minutes or until the top layer of the ester disappears. Record any changes in the odor of the ester. Remove the test tube and cool in cold water.

C.4 Write the equation for the saponification reaction.

C.5 After the solution is cool, add about 20 drops (1 mL) of 10% HCl until a drop of the solution turns blue litmus paper red. Record your observations. Determine the formula of the solid that forms.

Report Sheet - Lab 28

Date _____   Name _____

Section _____   Team _____

Instructor _____

Pre-Lab Study Questions

1. What are some carboxylic acids you encounter in daily life?

2. What are esters?

A.   Carboxylic Acids and Their Salts

	Acetic Acid	**Benzoic Acid**
A.1 Condensed structural formulas		
A.2 Solubility in cold water		
A.3 pH		
A.4 Solubility in hot water	✕	
A.5 NaOH		
A.6 HCl		

Write equations for the reactions of the carboxylic acids with NaOH (A.5).

Write equations for the reactions of the salts of the carboxylic acids with HCl (A.6).

Report Sheet - Lab 28

Questions and Problems

Q.1 How does NaOH affect the solubility of benzoic acid in water? Why?

Q.2 Write the names of the following carboxylic acids and esters:

a.

$$CH_3CH_2CH_2 - \overset{\overset{\displaystyle O}{\|}}{C} - OH$$

b.

$$CH_3CH_2\overset{\overset{\displaystyle O}{\|}}{C} - OCH_3$$

Q.3 Why are there differences in the solubility of the carboxylic acids in part A?

B. Esters

B.1 Equation for the formation of methyl acetate:

Report Sheet - Lab 28

B.2 Condensed Structural Formulas of Alcohol and Carboxylic Acid	B.3 Odor of Ester	Condensed Structural Formula and Name of Ester
Methanol and salicylic acid		
1-Pentanol and acetic acid		
1-Octanol and acetic acid		
Benzyl alcohol and acetic acid		
1-Propanol and acetic acid		

Report Sheet - Lab 28

C. Hydrolysis of Esters

C.1 Condensed structural formula of methyl salicylate	
C.2 Appearance and odor of methyl salicylate	
C.3 Describe the appearance and odor of the ester after adding NaOH and heating	
C.4 Write the equation for the saponification of the ester	
C.5 What changes occur when HCl is added?	
What is the formula of the compound that formed when HCl was added?	

Aspirin and Other Analgesics

Goals

- Use an esterification reaction to synthesize aspirin.
- Purify the crude aspirin sample.
- Test the purity of prepared aspirin and commercial aspirin products.
- Determine the physical and chemical properties of aspirin.
- Use thin-layer chromatography to separate and identify substances in analgesics.

Discussion

In the 18th century, an extract of willow bark was found useful in reducing fevers (antipyretic) and relieving pain and inflammation. Although salicylic acid was effective at reducing fever and pain, it damaged the mucous membranes of the mouth and esophagus, and caused hemorrhaging of the stomach lining. At the turn of the century, scientists at the Bayer Company in Germany noted that salicylic acid contained a phenol group that might cause the damage. They decided to modify the salicylic acid by forming an ester with a two-carbon acetyl group. The resulting substance was acetylsalicylic acid, or ASA, which we call aspirin. Aspirin acts by inhibiting the formation of prostaglandins, 20-carbon acids that form at the site of an injury and cause inflammation and pain.

In commercial aspirin products, a small amount of acetylsalicylic acid (300 mg to 400 mg) is bound together with a starch binder and sometimes caffeine and buffers to make an aspirin tablet. The basic conditions in the small intestine break down the acetylsalicylic acid to yield salicylic acid, which is absorbed into the bloodstream. The addition of a buffer reduces the irritation caused by the carboxylic acid group of the aspirin molecule.

A. Preparation of Aspirin

Aspirin (acetylsalicylic acid) can be prepared from acetic acid and the hydroxyl group on salicylic acid. However, this is a slow reaction. The ester forms rapidly when acetic anhydride is used to provide the acetyl group. *The aspirin you will prepare in this experiment is impure and must not be taken internally!*

Salicylic acid (138 g/mole) + Acetic anhydride → Aspirin (acetylsalicylic acid) (180 g/mole) + Acetic acid

Using the following equation, the maximum amount (yield) of aspirin that is possible from 2.00 g of salicylic acid can be calculated.

$$2.00 \text{ g salicylic acid} \times \frac{1 \text{ mole salicylic acid}}{138 \text{ g}} \times \frac{1 \text{ mole aspirin}}{1 \text{ mole salicylic acid}} \times \frac{180 \text{ g}}{1 \text{ mole aspirin}}$$

$$= 2.61 \text{ g aspirin (possible)}$$

Suppose the total amount of aspirin you obtain has a mass of 2.25 g. A percentage yield can be calculated as follows:

$$\% \text{ Yield} = \frac{\text{g aspirin obtained}}{\text{g aspirin calculated}} \times 100 = 86.2\% \text{ yield of aspirin product}$$

B.   Testing Aspirin Products

The purity of the crude sample and the recrystallized aspirin product can be tested with ferric chloride, $FeCl_3$. The Fe^{3+} ion reacts with the phenol group on salicylic acid and gives a purple color. This test can also be used to determine the purity of commercially prepared aspirin. Sometimes old aspirin breaks down to give salicylic acid and acetic acid. Then the aspirin in the bottle smells like vinegar and should be discarded.

C.   Analysis of Analgesics

Aspirin is one of several analgesics that are used to relieve pain. Other analgesics include acetaminophen, ibuprofen, and naproxen. Many aspirin products include caffeine. These products including aspirin are used to reduce fever, which means they are also antipyretics. However, aspirin also has anti-inflammatory properties and may reduce the risk of a heart attack.

Aspirin   **Acetaminophen**   **Ibuprofen**

Naproxen   **Caffeine**

Thin-Layer Chromatography (TLC)

Thin-layer chromatography (TLC) is a technique used to separate substances in a mixture. A TLC plate is typically a sheet of plastic, coated with a thin layer of a solid adsorbent such as silica gel.

Small amounts of known and unknown substances are placed as small spots at one end of the TLC plate. Then the end of the plate with the spots is placed in a solvent contained in a developing chamber.

The solid silica layer on the TLC plate is called the *stationary phase*. The solvent or the *moving phase* slowly moves up the silica layer on the TLC plate carrying the substances in the spot with it. The more soluble a substance is in the solvent, the higher the solvent will carry it up the plate. A substance that adheres strongly to the stationary silica gel moves only a short distance with the solvent. Thus, differences in the substances determines the distances they travel up the plate.

As the solvent front nears the top of the TLC plate, the plate is removed, marked, and dried. Then the substances are visualized. If they have colors, they can be seen directly. In this experiment, they are colorless. Because the silica material on the plate contains a fluorescent compound, ultraviolet light (254-nm) from a UV lamp can be used to visualize the substances, which appear as dark spots on the plate.

Calculating R_f Values

A value called the R_f value can be calculated for each substance on a plate. The R_f is the distance that a substance moves on the plate divided by the distance the solvent moves. An unknown substance is identified if its R_f value matches the R_f value of one of the known substances used on the plate. (See Figure 29.1.)

$$R_f = \frac{\text{distance substance moves}}{\text{distance solvent moves}}$$

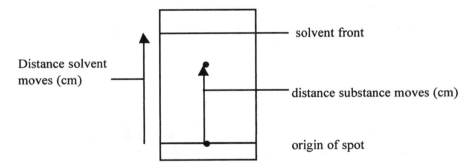

Figure 29.1 Distances moved by a substance and solvent on a TLC plate

In this experiment, you will use TLC to determine the R_f values for several known analgesics. You will also determine the R_f values and identify the types of analgesics in a variety of over the counter drugs used to relieve pain.

Experimental Procedures

WEAR PROTECTIVE GOGGLES AT ALL TIMES!

A. Preparation of Aspirin

Materials: 125-mL Erlenmeyer flask, 400-mL beaker, hot plate or Bunsen burner, ice, salicylic acid, acetic anhydride, 5- or 10-mL graduated cylinder, stirring rod, pan or large beaker, dropper, 85% H_3PO_4 in a dropper bottle, Büchner filtration apparatus, filter paper, spatula, watch glass

A.1 Weigh a 125-mL Erlenmeyer flask. Add 2.00 g of salicylic acid and reweigh. *Working in the hood, carefully* add 5 mL of acetic anhydride to the flask.

Caution: Acetic anhydride is irritating to the nose and sinus. Handle carefully.

Slowly add 10 drops of 85% phosphoric acid, H_3PO_4. Stir the mixture with a stirring rod. Place the flask and its contents in a boiling water bath and stir until all the solid dissolves.

Remove the flask from the hot water and let it cool to room temperature. *Working in the hood, cautiously* add 20 drops of water to the cooled mixture.

> ### *KEEP YOUR FACE AWAY FROM THE TOP OF THE FLASK:*
> ### *ACETIC ACID VAPORS ARE IRRITATING.*

When the reaction is complete, add 50 mL of cold water. Cool the mixture by placing the flask in an ice bath for 10 minutes. Stir. Crystals of aspirin should form. If no crystals appear, gently scratch the sides of the flask with a stirring rod.

Collecting the Aspirin Crystals

Some Büchner filtration apparatuses should be set up in the lab. Add a piece of filter paper. Place the funnel in the filter flask making sure that the neck fits snugly in a rubber washer. Moisten the filter paper. Turn on the water aspirator and pour the aspirin product onto the filter paper in the Büchner funnel. Push down gently on the funnel to create the suction needed to pull the water off the aspirin product. The aspirin crystals will collect on the filter paper. See Figure 29.2.

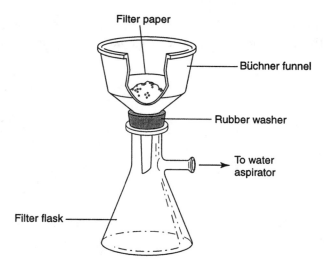

Figure 29.2 Apparatus for suction filtration with a Büchner funnel

Use a spatula to transfer any crystals left in the flask. Rinse the inside of the flask with a 10-mL portion of ice cold water to transfer all the crystals to the funnel. Wash the aspirin crystals on the filter paper with two 10-mL portions of cold water.

Spread the aspirin crystals out on the filter paper and draw air through the funnel. This helps dry the crystals. Turn off the water and use a spatula to lift and transfer the filter paper and aspirin to a paper towel. Don't touch it; it may still contain acid. Allow the crystals to air dry.

A.2 Weigh a clean, dry watch glass. Transfer the aspirin crystals to the watch glass and reweigh.

Calculations

A.3 Calculate the mass of salicylic acid.

A.4 Calculate the maximum yield of aspirin possible from the salicylic acid.

A.5 Calculate the mass of the aspirin you collected.

A.6 Calculate the percent yield of aspirin.

A.7 If a melting point apparatus is available, determine the melting point of your aspirin product. Pure aspirin has a melting point of 135°C. Salicylic acid melts at 157–159°C. Compare the melting point of your aspirin with the known melting points of aspirin and salicylic acid.

B. Testing Aspirin Products

Materials: Test tubes, spatula, aspirin from part A, commercial aspirin tablets, buffered aspirin, acetylsalicylic acid, 0.15% (m/v) salicylic acid, pH indicator paper, stirring rod, 1% $FeCl_3$, 10% NaOH, 10% HCl, 400-mL beaker, hot plate or Bunsen burner, blue litmus paper

B.1 **pH of aspirin** Place 3 mL of 0.15% salicylic acid in the first test tube. In test tubes 2–5, place a few crystals (the amount on the tip of a spatula) of the following substances and add 3 mL of water to each:

1. 0.15% Salicylic acid
2. Commercial aspirin (crushed)
3. Buffered aspirin (crushed)
4. Aspirin product from part A
5. Acetylsalicylic acid

Stir each mixture and touch the stirring rod to a piece of pH indicator paper. Compare the color of the paper to the chart on the container. Record the pH of each. *Save these test tubes and samples for part B.2.*

B.2 **Testing aspirin purity** To each of the samples from B.1, add 5 drops of 1% ferric chloride ($FeCl_3$) solution. Any free salicylic acid (unreacted during synthesis or resulting from hydrolysis in the breakdown of aspirin) reacts with the $FeCl_3$ to give a purple color. The more salicylic acid in the sample, the deeper the color. This indicates that the product is impure or that decomposition has taken place.

The maximum salicylic acid allowed in commercially prepared aspirin products is 0.15%. If the sample test has a lighter color than a 0.15% standard, the sample would be considered *pure* by USP standards. If the sample is darker, it is *impure* and not safe for ingestion. However, no matter what the results of the test, your laboratory-prepared aspirin must not be ingested. Record the colors. Compare the purity of the tested products to the reference sample of salicylic acid.

C.   Analysis of Analgesics

Materials: 400-mL beaker (developing chamber), Saran wrap, rubber band to fit beaker, solvent (75% ethyl acetate and 25% hexane), TLC plate with silica gel, UV lamp (short wave 254 nm), micropipettes, spot plates, dropper bottle containing 1% solutions in ethanol of aspirin, ibuprofen, acetaminophen, naproxen, caffeine, over the counter drugs, ruler

Preparing the TLC Developing Chamber

Obtain a 400-mL beaker, a piece of Saran wrap that covers, and a rubber band. Carefully pour a small amount of solvent into the beaker to a level of 0.5 – 0.6 cm. *It is important that the solvent level is below the spots you place on the TLC plate.* Cover the beaker with Saran wrap and secure with a rubber band.

Spotting the TLC Plate

Obtain a TLC plate that is 6 cm × 10 cm. *Be sure you handle the plates at the edge only to avoid transferring substances from your fingers.* Draw a light line with pencil about 1 cm above the end. This is your starting line or *origin*. Mark 6 dots on the line equally spaced. Label the dots from 1 through 6.

Place a few drops of each of the 1% solutions in a spot plate. Number the wells as follows.
  1. aspirin 2. ibuprofen   3. acetaminophen
  4. naproxen   5. caffeine   6. over the counter drug

Using clean capillary pipettes, one for each substance, spot a tiny amount of each substance on a dot. Lightly tap the micropipette to deliver a small amount. When dry you can apply again. The spot must be kept small rather than allowed to flow into larger spots. (See Figure 29.3.)

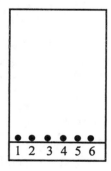

Figure 29.3   Spots of analgesics on a TLC plate

Placing TLC Plate in Developing Chamber

Carefully set the plate in the solvent in the beaker you prepared as the developing chamber. *The solvent must be lower than the origin on the plate.* Cover the beaker with Saran. Allow the beaker to remain undisturbed as the solvent moves up the TLC plate. (See Figure 29.4.)

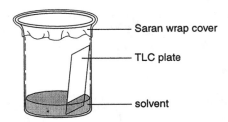

Figure 29.4 A developing chamber containing TLC plate.

When the solvent has risen almost to the top of the plate, open the chamber and draw a pencil line along the solvent front. Remove the plate and allow the solvent to evaporate in the hood. Place used solvent in the organic solvent container.

C.1 Observe the TLC place under UV light. Circle each spot. Draw a picture of the spots on your TLC plate.

C.2 Measure the distance from the origin to the solvent front. Measure the distance from the origin to the center of each spot.

C.3 Calculate the R_f value for each analgesic.

C.4 Identify the analgesics in the over the counter pain reliever.

Report Sheet - Lab 29

Date _____   Name _____

Section _____   Team _____

Instructor _____

Pre-Lab Study Questions

1. What functional group of aspirin causes it to irritate the stomach?

2. Why are buffers added to some aspirin products?

3. What quantity of aspirin is contained in most over-the-counter aspirin products?

Report Sheet - Lab 29

A. Preparation of Aspirin

A.1 Mass of flask   _____

     Mass of flask and salicylic acid   _____

A.2 Mass of watch glass   _____

Mass of watch glass and crude aspirin product   _____

Calculations

A.3 Mass of salicyclic acid   _____

A.4 Possible (maximum) yield of aspirin   _____

    (*Show calculations.*)

A.5 Mass of aspirin   _____

A.6 Percent yield   _____

    (*Show calculations.*)

A.7 Melting point (°C) of aspirin product (optional)   _____

Questions and Problems

Q.1 Write the structural formula for aspirin. Label the ester group and the carboxylic acid group.

Q.2 If a typical aspirin tablet contains 325 mg aspirin (the rest is starch binder), how many tablets could you prepare from the aspirin you made in lab?

Report Sheet - Lab 29

B. Testing Aspirin Products

Samples Tested	B.1 pH	B.2 Color with $FeCl_3$
1. 0.15% Salicylic acid		
2. Commercial aspirin brand:		
3. Buffered aspirin brand:		
4. Aspirin from A		
5. Acetylsalicylic acid		

Questions and Problems

Q.3 Give an explanation for any differences in the pH values in the samples you tested in part B.1.

Q.4 How does the pH of buffered aspirin product compare to the pH of the nonbuffered aspirin product?

Q.5 What substance is present if the $FeCl_3$ test gives a purple color? Which sample is the most impure?

Q.6 Aspirin that has been stored for a long time may give a vinegar odor and give a purple color with $FeCl_3$. What reaction would cause this to happen?

Report Sheet - Lab 29

C. Analysis of Analgesics

C.1

C.2 Distance moved by solvent _____ cm

Spot #	Analgesic substance	C.2 Distance moved	C.3 R_f value
1			
2			
3			
4			
5			
6			

C.4 Substance(s) present in over the counter pain reliever
According to your R_f values, what substance(s) can you identify as present in the over the counter pain reliever in your analysis?

Substance 1 _____

Substance 2 _____

Goals

- Observe the physical and chemical properties of some common lipids.
- Draw the structure of a typical triacylglycerol.
- Distinguish between saturated and unsaturated fats.
- Determine the degree of unsaturation of some fats.
- Prepare a hand lotion and determine the function of its components.

Discussion

A. Triacylglycerols

The triacylglycerols, commonly called fats or oils, are esters of glycerol and fatty acids. Fatty acids are long-chain carboxylic acids, usually 14 to 18 carbons in length. When the fatty acid contains double bonds, the triacylglycerol is referred to as an unsaturated fat. When the fatty acid consists of an alkane-like carbon chain, the triacylglycerol is a saturated fat. Table 30.1 gives the formulas of the common fatty acids and their melting points. At room temperature, saturated fats are usually solid and unsaturated fats are usually liquid.

Table 30.1 *Formulas, Melting Points, and Sources of Some Fatty Acids*

Carbon Atoms	Structural Formula	Melting Point (°C)	Common Name	Source
Saturated fatty acids (single carbon–carbon bonds)				
12	$CH_3(CH_2)_{10}COOH$	44	lauric	coconut
14	$CH_3(CH_2)_{12}COOH$	54	myristic	nutmeg
16	$CH_3(CH_2)_{14}COOH$	63	palmitic	palm
18	$CH_3(CH_2)_{16}COOH$	70	stearic	animal fat
Monounsaturated fatty acids (one cis double bond)				
16	$CH_3(CH_2)_5CH=CH(CH_2)_7COOH$	1	palmitoleic	butter fat
18	$CH_3(CH_2)_7CH=CH(CH_2)_7COOH$	4	oleic	olives, corn
Polyunsaturated fatty acids (two or more cis double bonds)				
18	$CH_3(CH_2)_4CH=CHCH_2CH=CH(CH_2)_7COOH$	-5	linoleic	safflower, sunflower
18	$CH_3CH_2CH=CHCH_2CH=CHCH_2CH=CH(CH_2)_7COOH$	-11	linolenic	corn

Fats that contain mostly saturated fatty acids have a higher melting point than fats with more unsaturated fatty acids.

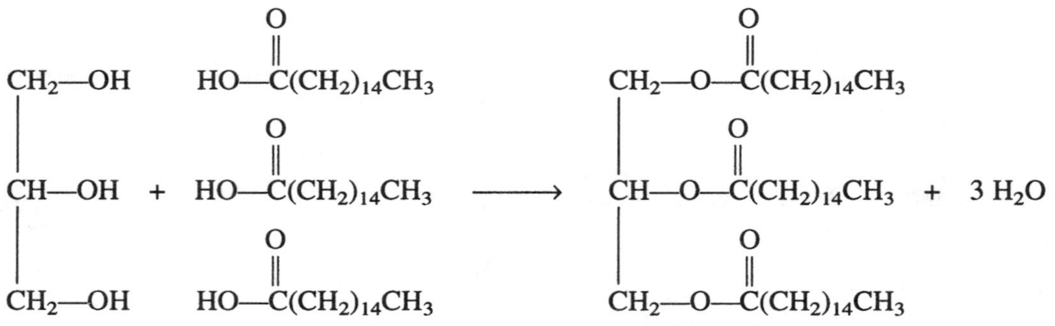

$$
\begin{array}{llll}
\text{CH}_2\text{—OH} & \text{HO—}\overset{\displaystyle O}{\overset{\|}{\text{C}}}\text{(CH}_2)_{14}\text{CH}_3 & & \text{CH}_2\text{—O—}\overset{\displaystyle O}{\overset{\|}{\text{C}}}\text{(CH}_2)_{14}\text{CH}_3 \\
\text{CH—OH} \;+\; & \text{HO—}\overset{\displaystyle O}{\overset{\|}{\text{C}}}\text{(CH}_2)_{14}\text{CH}_3 & \longrightarrow & \text{CH—O—}\overset{\displaystyle O}{\overset{\|}{\text{C}}}\text{(CH}_2)_{14}\text{CH}_3 \;+\; 3\,\text{H}_2\text{O} \\
\text{CH}_2\text{—OH} & \text{HO—}\overset{\displaystyle O}{\overset{\|}{\text{C}}}\text{(CH}_2)_{14}\text{CH}_3 & & \text{CH}_2\text{—O—}\overset{\displaystyle O}{\overset{\|}{\text{C}}}\text{(CH}_2)_{14}\text{CH}_3
\end{array}
$$

Glycerol   3 Palmitic acids   Glyceryl palmitate (tripalmitin)

B. Physical Properties of Some Lipids and Fatty Acids

Lipids are a family of compounds that are grouped by similarities in solubility rather than structure. As a group, lipids are more soluble in nonpolar solvents such as ether, chloroform, or benzene. Most are not soluble in water. Important types of lipids include fats and oils, glycerophospholipids, and steroids. Compounds classified as lipids include fat-soluble vitamins A, D, E, and K; cholesterol; hormones; portions of cell membranes; and vegetable oils. Table 30.2 lists the classes of lipids.

Table 30.2 *Classes of Lipid Molecules*

Lipids	Components
Waxes	Fatty acid and long-chain alcohol
Fats and oils (triacylglycerols)	Fatty acids and glycerol
Glycerophospholipids	Fatty acids, glycerol, phosphate, amino alcohol
Sphingolipids	Fatty acids, sphingosine, phosphate, amino alcohol
Glycosphinolipids	Fatty acids, glycerophospholipids sphingosine, monosaccharides
Steroids	A fused structure of three cyclohexanes and a cyclopentane

The structural formulas of three typical lipids are shown below:

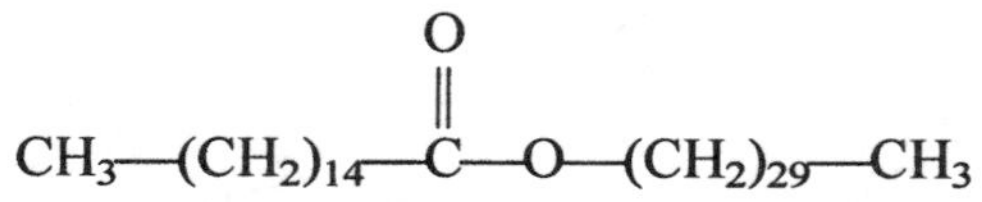

$$\text{CH}_3\text{—(CH}_2)_{14}\text{—}\overset{\displaystyle O}{\overset{\|}{\text{C}}}\text{—O—(CH}_2)_{29}\text{—CH}_3$$

Wax

$$
\begin{array}{l}
\text{CH}_2\text{—O—}\overset{\displaystyle O}{\overset{\|}{\text{C}}}\text{—(CH}_2)_{16}\text{—CH}_3 \\
\text{CH—O—}\overset{\displaystyle O}{\overset{\|}{\text{C}}}\text{—(CH}_2)_{16}\text{—CH}_3 \\
\text{CH}_2\text{—O—}\overset{\displaystyle O}{\overset{\|}{\text{C}}}\text{—(CH}_2)_{16}\text{—CH}_3
\end{array}
$$

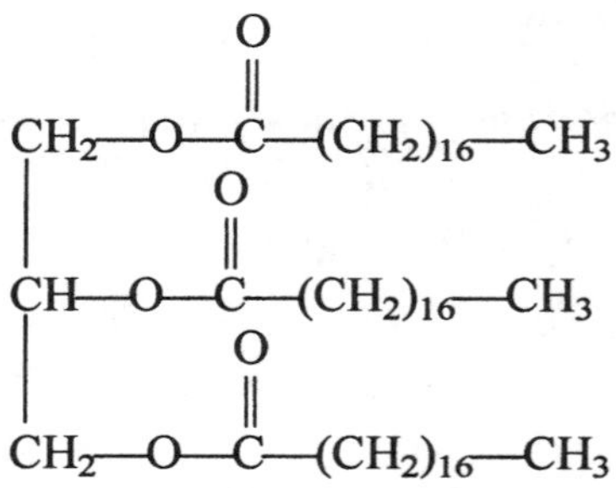

Triacylglycerol, a fat

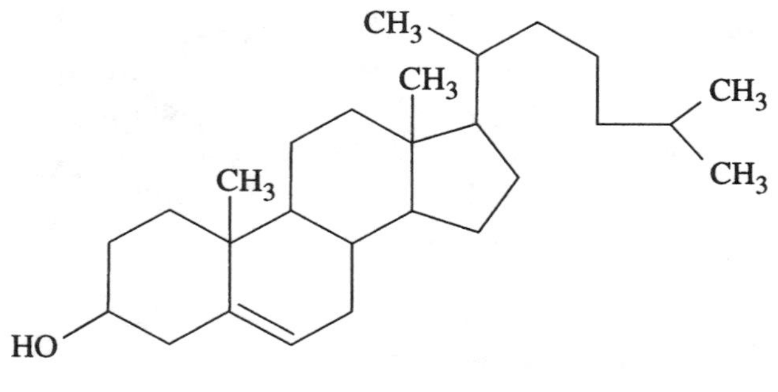

Cholesterol, a steroid

C. Bromine Test for Unsaturation

The presence of unsaturation in a fatty acid or a triacylglycerol can be detected by the bromine test, which you used in an earlier experiment to detect double bonds in alkenes. If the orange color of the bromine solution fades quickly, an addition reaction has occurred and the oil or fat is unsaturated.

Bromine adds to the double bond

$$CH_2-O-\overset{\overset{\displaystyle O}{\|}}{C}(CH_2)_7CH=CH(CH_2)_7CH_3$$
$$CH-O-\overset{\overset{\displaystyle O}{\|}}{C}(CH_2)_{16}CH_3$$
$$CH_2-O-\overset{\overset{\displaystyle O}{\|}}{C}(CH_2)_{16}CH_3$$

$+ Br_2 \longrightarrow$

$$CH_2-O-\overset{\overset{\displaystyle O}{\|}}{C}(CH_2)_7\overset{\overset{\displaystyle Br}{|}}{CH}-\overset{\overset{\displaystyle Br}{|}}{CH}(CH_2)_7CH_3$$
$$CH-O-\overset{\overset{\displaystyle O}{\|}}{C}(CH_2)_{16}CH_3$$
$$CH_2-O-\overset{\overset{\displaystyle O}{\|}}{C}(CH_2)_{16}CH_3$$

D. Preparation of Hand Lotion

We use hand lotions and creams to soften our skin and reduce dryness. Typically, the formulation of a hand lotion consists of several components such as stearic acid, lanolin, triethanolamine, cetyl alcohol, glycerin (glycerol), water, and usually a fragrance. Lanolin from wool consists of a mixture of waxes.

Cetyl alcohol	$CH_3(CH_2)_{15}OH$
Stearic acid	$CH_3-(CH_2)_{16}-\overset{\overset{\displaystyle O}{\|}}{C}-OH$
Glycerol (glycerine)	CH_2-OH $\|$ $CH-OH$ $\|$ CH_2-OH
Triethanolamine	$\overset{\displaystyle CH_2CH_2OH}{\underset{\displaystyle |}{}}$ $HOCH_2CH_2-N-CH_2CH_2OH$

Because lipids are nonpolar, they protect and soften by preventing the loss of moisture from the skin. Some of the components help emulsify the polar and nonpolar ingredients. In this experiment, we will see how the physical and chemical properties of lipids are used to prepare a hand lotion.

Lab Information

Time: 3 hr

Comments: Bromine can cause severe chemical burns. Use carefully.
                Tear out the Lab report sheets and place them beside the matching procedures.

Related topics: Fatty acids, saturated and unsaturated fatty acids, lipids, triglycerides

Experimental Procedures

BE SURE YOU ARE WEARING YOUR SAFETY GOGGLES!

A. Triacylglycerols

Materials: Organic model kits or models

A.1 Use an organic model kit or observe prepared models of a molecule of glycerol and three molecules of ethanoic acid. What are the functional groups on each? Draw their structures.

A.2 Form ester bonds between the hydroxy groups on glycerol and the carboxylic acid groups of the ethanoic acid molecules. In the process, three molecules of water are removed. Write an equation for the formation of the glyceryl ethanoate.

Carry out the reverse process, which is hydrolysis. Add the components of water to break the ester bond. Add an arrow to the equation to show the reverse direction for the hydrolysis reaction.

B. Physical Properties of Some Lipids and Fatty Acids

Materials: Test tubes and stoppers, dropper bottles or solids: stearic acid, oleic acid, olive oil, safflower oil, lecithin, cholesterol, vitamin A capsules, spatulas, CH_2Cl_2 (optional)

To seven separate test tubes, add 5 drops or the amount of solid lipid held on the tip of a spatula: stearic acid, oleic acid, olive oil, safflower oil, lecithin, cholesterol, vitamin A (puncture a capsule or use cod liver oil).

Appearance and Odor

B.1 Classify each as a triacylglycerol (fat or oil), fatty acid, steroid, or phospholipid.

B.2 Describe their appearance.

B.3 Describe their odors.

Solubility in a Polar Solvent *(May be a demonstration)*

B.4 Add about 2 mL of water to each of the test tubes. Stopper and shake each test tube. Record your observations.

Solubility in a Nonpolar Solvent *(Optional or demonstration)*

B.5 Place 5 drops, or a small amount of solid, of the following in separate test tubes: stearic acid, oleic oil, olive oil, safflower oil, lecithin, cholesterol, and vitamin A. Add 1 mL (20 drops) methylene chloride, CH_2Cl_2, to each sample. Record the solubility of the lipids. *Save the test tubes and samples of stearic acid oleic acid, olive oil and safflower oil for part C.*

C. Bromine Test for Unsaturation

Materials: Samples from B.5, 1% Br_2 in methylene chloride

To the samples from B.5, add 1% bromine solution drop by drop until a permanent red-orange color is obtained or until 20 drops have been added.

Caution: Avoid contact with bromine solution; it can cause painful burns. Do not breathe the fumes.

Record your observations. Determine if the red-orange color fades rapidly or persists.

D. Preparation of Hand Lotion

Materials: Stearic acid, cetyl alcohol, lanolin (anhydrous), triethanolamine, glycerol, ethanol, distilled water, fragrance (optional), commercial hand lotions, 10-mL graduated cylinder, 50-mL graduated cylinder, 50-mL or 100-mL beakers, thermometer, 100-mL beakers, 250-mL beaker for water bath, Bunsen burner, iron ring, wire screen, stirring rods, tongs, pH paper

Team project: D.1, D.2, and D.3 may be prepared by different teams in the lab.

D.1 Obtain the following substances and combine in two 50-mL or 100-mL beakers. Use a laboratory balance to weigh out the solid substances. Use a 10-mL graduated cylinder to measure small volumes, and a 50-mL graduated cylinder to measure larger volumes.

Beaker 1		**Beaker 2**	
Stearic acid	3 g	Glycerin	2 mL
Cetyl alcohol	1 g	Water	50 mL
Lanolin (anhydrous)	2 g		
Triethanolamine	1 mL		

Water bath: Fill a 250-mL beaker about 2/3 full of water. Place the beaker on an iron ring covered with a wire screen. Lower a second iron ring that fits around the 250-mL beaker to stabilize it. Turn on the Bunsen burner to heat the water.

Using a pair of crucible tongs, hold Beaker 1 (four ingredients) in the water bath and heat to 80°C or until all the compounds have melted. Remove.

Using a pair of crucible tongs, place Beaker 2 (two ingredients) in the same water bath and heat to 80°C.

While still warm, slowly pour the glycerol-water mixture from Beaker 2 into Beaker 1 (four ingredients) as you stir. Add 5.0 mL of ethanol and a few drops of fragrance, if desired. Continue to stir for 3-5 minutes until a smooth, creamy lotion is obtained. If the resulting product is too thick, add more warm water.

Describe the smoothness and appearance of the hand lotion.

D.2 Repeat the experiment, but omit the triethanolamine from Beaker 1. Compare the properties of the resulting hand lotion to the one obtained in D.1 and to commercial hand lotions.

D.3 Repeat the experiment, but omit the stearic acid from Beaker 1. Compare the properties and textures of the resulting hand lotion to the one obtained in D.1 and to commercial hand lotions.

D.4 Determine the pH of the hand lotions you and others in your lab have prepared along with any commercial hand lotions.

Report Sheet - Lab 30

Date _____   Name _____

Section _____   Team _____

Instructor _____

Pre-Lab Study Questions

1. What is the functional group in a triacylglycerol?

2. Write the structure of linolenic acid. Why is it an unsaturated fatty acid?

A.   Triacylglycerols

A.1   Structure of glycerol   Structure of ethanoic acid

A.2   Reversible equation for the esterification and hydrolysis of glyceryl ethanoate

Report Sheet - Lab 30

B. Physical Properties of Some Lipids and Fatty Acids

Lipid	B.1 Type of Lipid	B.2 Appearance	B.3 Odor	B.4 Soluble in Water?	B.5 Soluble in CH_2Cl_2?
Stearic acid					
Oleic acid					
Olive oil					
Safflower oil					
Lecithin					
Cholesterol					
Vitamin A					

Questions and Problems

Q.1 Why are the compounds in part B classified as lipids?

Q.2 What type of solvent is needed to remove an oil spot? Why?

Report Sheet - Lab 30

C. Bromine Test for Unsaturation

Compound	Drops of Bromine Solution	Saturated or Unsaturated?
Fatty acids		
Stearic acid		
Oleic acid		
Triacylglycerols		
Safflower oil		
Olive oil		

Questions and Problems

Q.3 a. Write the condensed structural formulas of stearic acid and oleic acid.

Stearic acid

Oleic acid

b. Which fatty acid is unsaturated?

c. The melting point of stearic acid is 70°C, and that of oleic acid is 4°C. Explain the difference.

d. From the results of experiment C, how can you tell which is more unsaturated, oleic acid or stearic acid?

D. Preparation of Hand Lotion

Descriptions of the hand lotions		
D.1	D.2 (without triethanolamine)	D.3 (without stearic acid)
D.4 pH of the hand lotions		
pH of commercial hand lotions		
Brand_____	Brand_____	Brand_____
pH _____	pH _____	pH _____

Questions and Problems

Q. 4   How does omitting triethanolamine affect the properties and appearance of the hand lotion?

Q. 5   How does omitting stearic acid affect the properties and appearance of the hand lotion?

Q. 6   What would be a reason to have triethanolamine and stearic acid as ingredients in hand lotion?

Glycerophospholipids and Steroids

Goals

- Observe the physical and chemical properties of some common lipids.

- Draw the structure of a typical glycerophospholipid.

- Draw the structure of cholesterol.

- Isolate lecithin and cholesterol from egg yolk.

Egg yolks contain about 10% lecithin, a glycerophospholipid, as well as cholesterol, a steroid. Other sources of lecithin are soybeans, peanuts, beef, and cauliflower. Lecithin contains two fatty acids (nonpolar) along with the ester of the amino alcohol choline (polar). The fatty acids found in lecithin are palmitic 11.7%, stearic 4.0%, palmitoleic 8.6%, oleic 9.8%, linoleic 55.0%, and linolenic 4.0%. Because lecithin consists of both polar and nonpolar sections, it is used to stabilize foods such as mayonnaise, salad dressings and Hollandaise sauce. Cholesterol is a type of lipid that has a steroid ring structure and used in the synthesis of several hormones

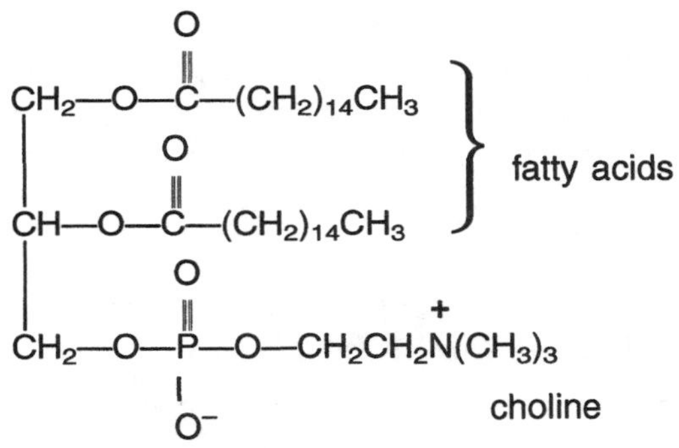

Lecithin, a glycerophospholipid

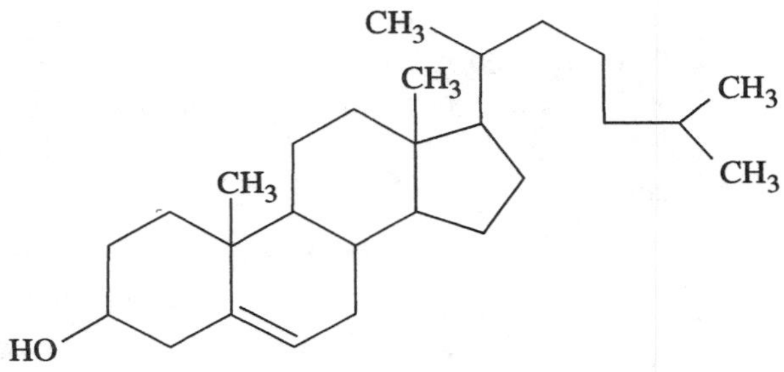

Cholesterol, a steroid

Lecithin and cholesterol are extracted from egg yolk using the organic solvents acetone and ethyl ether. Because cholesterol is soluble in acetone and lecithin is not, acetone is used to extract cholesterol from the egg yolk material. Then ethyl ether is used to extract lecithin from the remaining yolk material.

Lab Information

Time: 2-3 hr

Comments: Acetone and ethyl ether are *highly flammable*. Work with these solvents in the hood. No flames are allowed in lab during this experiment.
Tear out the Lab report sheets and place them beside the matching procedures.

Related topics: Fatty acids, glycerophospholipids, steroids

Experimental Procedures

BE SURE YOU ARE WEARING YOUR SAFETY GOGGLES!

A. Isolating Cholesterol from Egg Yolk

Materials: 1 hard boiled egg, 250-mL beaker, acetone, 125-mL Erlenmeyer flask, 50-mL flask, steam bath, 100-mL beaker, spatula, short-stem funnel, glass wool, stirring rods, melting point apparatus

---

Caution: No open flames are allowed in lab during this experiment. Organic solvents are flammable!

A.1 Weigh a 250-mL beaker. Obtain an egg, peel it, and discard the egg white. Place the yolk in the beaker and break it up with a spatula or stirring rod. Determine the combined mass of the beaker and egg yolk.

Separating Cholesterol from Egg Yolk

Add 50 mL of acetone to the egg yolk in the 250-mL beaker. Stir the mixture periodically for the next 10-15 minutes. Allow the solid to settle to the bottom of the beaker. Slowly pour the acetone layer into a 125-mL Erlenmeyer flask. This acetone layer contains the cholesterol from the egg yolk. The solid material remaining in the beaker contains the lecithin. Place this beaker in a hood to allow the residual acetone to evaporate from the solid material. *Save this beaker and the solid material for the isolation of lecithin in part B.*

Carry out this part of the lab in the hood!

A.2 Place a small amount of glass wool in a short-stem funnel. Pour the acetone layer containing the cholesterol in the 125-mL Erlenmeyer flask through the glass wool into a 100-ml beaker. Place the beaker on a steam bath to reduce the volume of the extract to between 5 to 10 mL. You can place 5-10 mL in a matching beaker and mark the level to determine the level for the acetone mixture. *Do not let it evaporate to dryness.*

Weigh a 50-mL Erlenmeyer flask. Pour the reduced volume of acetone extract containing cholesterol into the flask. Cork and place the flask in an ice bath for at least 20 minutes. A white solid, which is crude cholesterol, should form. *While you are waiting for the cholesterol crystals to appear, proceed with the isolation of lecithin in B.*

After white solid has formed, remove the flask containing the cholesterol from the ice bath and pour off the cold acetone. Place the acetone in a container marked "waste acetone".

A.3 Set the flask in the hood and allow residual acetone to evaporate. When dry, weigh the flask and cholesterol crystals. Record.

A.4 Describe the color and appearance of cholesterol.

A.5 (*optional*) Determine the melting point of your crude cholesterol. The known melting point for cholesterol is 149°C.

Calculations

A.6 Calculate the mass of the egg yolk.

A.7 Calculate the mass of the cholesterol.

A.8 Determine the percent by mass of cholesterol in egg yolk.

$$\text{Mass \% cholesterol} = \frac{\text{g cholesterol}}{\text{g egg yolk}} \times 100$$

B. Isolating Lecithin from Egg Yolk

Materials: Egg yolk residue from Part A, ethyl ether, hot plate or steam bath, 100-mL beaker, stirring rods

B.1 Place the beaker from Part A containing the egg yolk residue on a steam bath at a low or moderate temperature to evaporate any remaining acetone without splattering. Remove if it starts to splatter. There will be only a small amount of acetone in the residue, so this takes only a few minutes. Remove the beaker and let it cool.

> **Caution: Ether is *extremely* flammable.
> Carry out this part of the lab in the hood!
> No open flames are allowed during this experiment.**

In the hood, add 30 mL of ethyl ether to the egg yolk residue to extract the lecithin. Stir strongly for 2-3 minutes.

Weigh a clean 100-mL beaker and record. Pour the ether extract, which contains the lecithin, into the weighed 100-mL beaker. Place the left over egg yolk material in a "waste" container as indicated by your instructor.

B.2 *In the hood,* place the beaker on a steam bath to evaporate the ether. The lecithin may appear as a thick syrup or waxy substance. Remove the beaker occasionally if the mixture starts to splatter. Cool and determine the combined mass of the beaker and lecithin.

B.3 Describe the color and appearance of lecithin. Oxidation may turn it to yellow or brown.

Calculations

B.4 Calculate the mass of the lecithin.

B.5 Determine the percent by mass of lecithin in egg yolk. You already have the mass of the egg yolk in A.6.

$$\text{Mass \% lecithin} = \frac{\text{g lecithin}}{\text{g egg yolk}} \times 100$$

Report Sheet - Lab 31

Date _____   Name _____

Section _____   Team _____

Instructor _____

Pre-Lab Study Questions

1. How are the structures of triacylglycerols similar to the structures of glycerophospholipids?

    How do they differ?

2. Draw the structure of the steroid nucleus.

A.   Isolation of Cholesterol from Egg Yolk

A.1 Mass of beaker and egg yolk   _____

     Mass of beaker   _____

A.2 Mass of flask   _____

A.3 Mass of flask and cholesterol   _____

A.4 Color and appearance of cholesterol   _____

A.5 Melting point of crude cholesterol   _____

Calculations

A.6 Mass of egg yolk   _____

A.7 Mass of the cholesterol   _____

A.8 Mass % of cholesterol in egg yolk   _____

    (*Show calculations*)

Report Sheet - Lab 31

Questions and Problems

Q.1 If a person has high blood cholesterol, why would a nutrionist recommend the use of egg white only in egg dishes?

Q.2 What groups are found on the steroid nucleus in the structure of cholesterol?

B. Isolation of Lecithin from Egg Yolk

B.1 Mass of beaker   _____

B.2 Mass of beaker and lecithin   _____

B.3 Color and appearance of lecithin

Calculations
B.4 Mass of the lecithin   _____

B.5 Mass % of lecithin in egg yolk   _____

   (*Show calculations*)

Questions and Problems

Q.3 Why would lecithin in egg yolk be more useful than a triacylglycerol in the emulsification of oil and water in the preparation of mayonnaise?

Q.4 Lecithins contain a high amount of linoleic acid. Draw the structure of a lecithin containing two linoleic fatty acids

   Linoleic acid $CH_3—(CH_2)_4—CH=CH—CH_2—CH=CH—(CH_2)—COOH$.

Saponification and Soaps

Goals

- Prepare soap by the saponification of a fat or oil.
- Observe the reactions of soap with oil, $CaCl_2$, $MgCl_2$, and $FeCl_3$.

Discussion

A. Saponification: Preparation of Soap

For centuries, soaps have been made from animal fats and lye (NaOH), which was obtained by pouring water through wood ashes. The hydrolysis of a fat or oil by a base such as NaOH is called *saponification* and the salts of the fatty acids obtained are called *soaps*. The other product of hydrolysis is glycerol, which is soluble in water.

The fats that are most commonly used to make soap are lard and tallow from animal fat and coconut, palm, and olive oils from vegetables. Castile soap is made from olive oil. Soaps that float have air pockets. Soft soaps are made with KOH instead of NaOH to give potassium salts.

B. Properties of Soaps and Detergents

A soap molecule has a dual nature. The nonpolar carbon chain is hydrophobic and attracted to nonpolar substances such as grease. The polar head of the carboxylate salt is hydrophilic and attracted to water.

When soap is added to a greasy substance, the hydrophobic tails are embedded in the non-polar fats and oils. However, the polar heads are attracted to the polar water molecules. Clusters of soap particles called *micelles* form with the nonpolar oil droplet in the center surrounded by many polar heads that extend into the water. Eventually all of the greasy substance forms micelles, which can be washed away with water. In hard water, the carboxylate ends of soap react with Ca^{2+}, Fe^{3+}, or Mg^{2+} ions and form an insoluble substance, which we see as a gray line in the bathtub or sink. Tests will be done with the soap you prepare to measure its pH, its ability to form suds in soft and hard water, and its reaction with oils.

Detergents or "syndets" are called synthetic cleaning agents because they are not derived from naturally occurring fats or oils. They are popular because they do not form insoluble salts with ions, which means they work in hard water as well as in soft water. A typical detergent is sodium lauryl sulfate.

$$CH_3(CH_2)_{10}CH_2 - O - \overset{\overset{\displaystyle O}{\|}}{\underset{\underset{\displaystyle O}{\|}}{S}} - O^-Na^+$$

Lauryl sulfate salt,
a nonbiodegradable detergent

As detergents replaced soaps for cleaning, it was found that they were not degraded in sewage treatment plants. Large amounts of foam appeared in streams and lakes that became polluted with detergents. Biodegradable detergents such as an alkylbenzenesulfonate detergent eventually replaced the nonbiodegradable detergents.

$$CH_3(CH_2)_9 - \overset{\overset{\displaystyle CH_3}{|}}{CH} - \underset{}{\bigcirc} - \overset{\overset{\displaystyle O}{\|}}{\underset{\underset{\displaystyle O}{\|}}{S}} - O^-Na^+$$

Laurylbenzenesulfonate salt,
a biodegradable detergent

In addition to the sulfonate salts, a box of detergent contains phosphate compounds along with brighteners and perfumes. However, phosphates accelerate the growth of algae in lakes and cause a decrease in the dissolved oxygen in the water. As a result, the lake decays. Some replacements for phosphates have been made.

Lab Information

Time: 2 hr
Comments: You will be working with hot oil and NaOH. Be sure you wear your goggles.
               Tear out the report sheets and place them beside the matching procedures.
Related Topics: Esters, saponification, soaps, hydrophobic, hydrophilic

Experimental Procedures

> ### *Wear your protective goggles!*

A. Saponification: Preparation of Soap

Materials: 150-mL beaker, hot plate, graduated cylinder, stirring rod or stirring hot plate with stirring bar, large watch glass, 400-mL beaker, Büchner filter system, filter paper, plastic gloves, fat (lard, solid shortening, coconut oil, olive or other vegetable oil), ethanol, 20% NaOH, saturated NaCl solution

Weigh a 150-mL beaker. Add about 5 g of fat or oil. Reweigh.

Add 15 mL ethanol (solvent) and 15 mL of 20% NaOH. *Use care when pouring NaOH.* Place the beaker on a hot plate and heat to a gentle boil and stir continuously. A magnetic stirring bar may be used with a magnetic stirrer. Heat for 30 minutes or until saponification is complete and the solution becomes clear with no separation of layers. Be careful of splattering; the mixture contains a strong base. Wear disposable gloves, if available. Do not let the mixture overheat or char. Add 5-mL portions of an ethanol–water (1:1) mixture to maintain volume. If foaming is excessive, *reduce* the heat.

Caution: Oil and ethanol will be hot, and may splatter or catch fire. Keep a watch glass nearby to smother any flames. NaOH is caustic and can cause permanent eye damage. Wear goggles at all times.

Obtain 50 mL of a saturated NaCl solution in a 400-mL beaker. (A saturated NaCl solution is prepared by mixing 30 g of NaCl with 100 mL of water.) Pour the soap solution into this salt solution and stir. This process, known as "salting out," causes the soap to separate out and float on the surface.

Collecting the soap Collect the solid soap using a Büchner funnel and filter paper. See Figure 32.1. Wash the soap with two 10-mL portions of cold water. Pull air through the product to dry it further. Place the soap curds on a watch glass or in a small beaker and dry the soap until the next lab session. Use disposable, plastic gloves to handle the soap. **Handle with care: The soap may still contain NaOH, which can irritate the skin.** Save the soap you prepared for the next part of this experiment. Describe the soap.

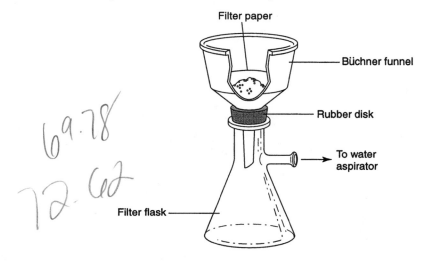

Figure 32.1 Apparatus for suction filtration with Büchner funnel

B. Properties of Soaps and Detergents

Materials: Test tubes, stoppers to fit, droppers, small beakers, 50- or 100-mL graduated cylinder, stirring rod, laboratory-prepared soap (from part A), commercial soap product, detergent, pH paper, oil, 1% $CaCl_2$, 1% $MgCl_2$, and 1% $FeCl_3$

Prepare solutions of the soap you made in part A, a commercial soap, and a detergent by dissolving 1 g of each in 50 mL of distilled water. If the soap is a liquid, use 20 drops.

B.1 **pH test** Place 10 mL of each soap solution in separate test tubes. Use 10 mL of water as a comparison. Label. Dip a stirring rod into each solution, then touch the stirring rod to pH paper. Determine the pH. *Save the tubes for part B.2.*

B.2 **Foam test** Stopper each of the tubes from B.1 and shake for 10 seconds. The soap should form a layer of suds or foam. Record your observations. *Save the tubes for part B.3.*

B.3 **Reaction with oil** Add 5 drops of oil to each test tube from B.2. Stopper and shake each one for 10 seconds. Record your observations. Compare the sudsy layer in each test tube to the sudsy layers in part B.2.

B.4 **Hard water test** Place 5 mL of the soap solutions in three separate test tubes. Add 20 drops of 1% $CaCl_2$ to the first sample, 20 drops of 1% $MgCl_2$ to the second tube, and 20 drops of 1% $FeCl_3$ to the third tube. Stopper each test tube and shake 10 seconds. Compare the foamy layer in each of the test tubes to the sudsy layer obtained in part B.2. Record your observations.

Report Sheet - Lab 32

Date ___4/10/06___ Name ___Lauren-Nicole Pascual___

Section ___CHEM-1301-AA___ Team _____

Instructor _____

Pre-Lab Study Questions

1. What happens when a fatty acid is reacted with NaOH?

   Potassium salts are created when a fatty acid reacts with NaOH

2. Why is ethanol added to the reaction mixture of fat and base in the making of soap?

   Since ethanol is an alcohol, ethanol is the key ingredient that makes soap anti-bacterial.

3. Why is the product of saponification a salt?

   It's base is NaOH.

A. Saponification: Preparation of Soap

Describe the appearance of your soap.

Off-white color, powdery yet clumped

Questions and Problems

Q.1 How would soaps made from vegetable oils differ from soaps made from animal fat?

Q.2 How does soap remove an oil spot?

B. Properties of Soaps and Detergents

Tests	Water	Lab Soap	Commercial Soap	Detergent
B.1 pH		pH 11	5	
B.2 Foam		yes	no	
B.3 Oil		yes		
B.4 CaCl$_2$		yes		
MgCl$_2$		yes		
FeCl$_3$				

Report Sheet - Lab 32

Questions and Problems

Q.3 Which of the solutions were basic? Why?

Q.4 Write an equation for the saponification of trimyristin with KOH.

Amines and Amides

Goals

- Draw the structural formulas and give the names of amines.
- Classify amines as primary, secondary, or tertiary.
- Observe some physical properties of amines and amides.
- Write an equation for the formation of an amine salt.
- Write an equation for the formation of an amide and its hydrolysis in acid and base.

Discussion

A. Structure and Classification of Amines

Amines are considered as derivatives of ammonia in which one or more hydrogen atoms are replaced with alkyl or aromatic groups. The number of alkyl groups attached to the nitrogen atom determines the classification of primary, secondary, or tertiary amines.

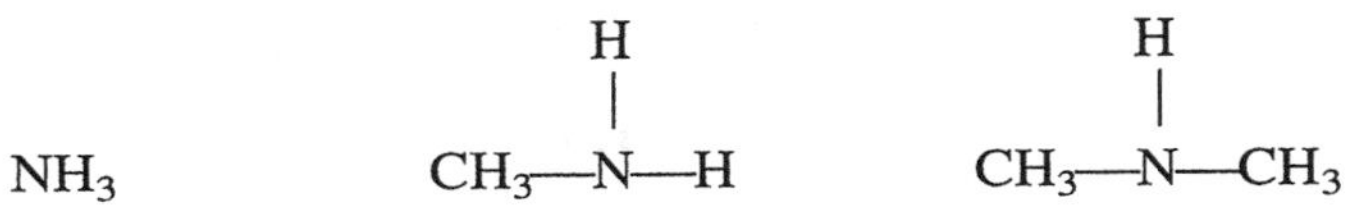

NH_3
Ammonia

Methylamine
(primary, 1°)

Dimethylamine
(secondary, 2°)

Trimethylamine
(tertiary, 3°)

Amines are often found as part of compounds that are physiologically active or used in medications.

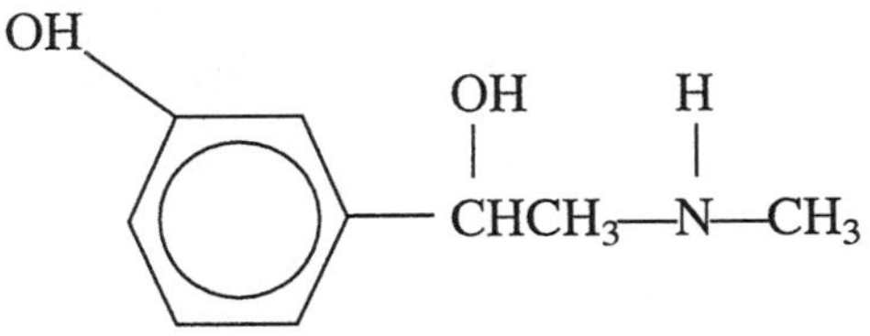

Neo-Synephrine

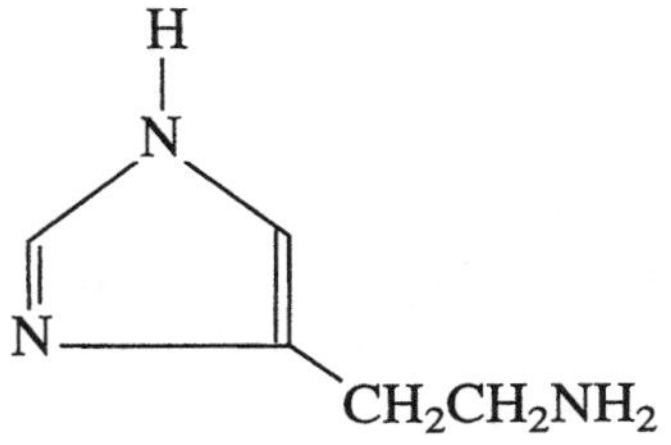

Histamine

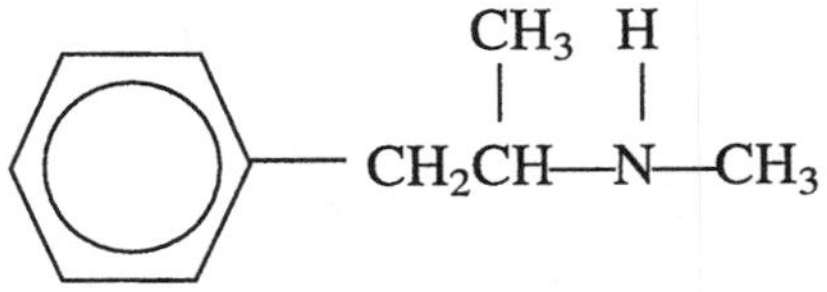

Methamphetamine (methedrine)

B. Solubility of Amines in Water

In water, ammonia and amines with one to four carbon atoms act as weak bases because the unshared pair of electrons on the nitrogen atom attracts protons. The products are an ammonium ion or alkyl ammonium ion and a hydroxide ion.

$$NH_3 \quad + \quad H_2O \longrightarrow \quad NH_4^+ \quad + \quad OH^-$$

Ammonia $\qquad\qquad\qquad$ Ammonium ion $\qquad$ Hydroxide ion

$$CH_3-NH_2 \quad + \quad H_2O \longrightarrow \quad CH_3-NH_3^+ \quad + \quad OH^-$$

Methylamine $\qquad\qquad\qquad$ Methylammonium ion $\qquad$ Hydroxide ion

C. Neutralization of Amines with Acids

Because amines are basic, they react with acids to form the amine salt. These amine salts are much more soluble in water than the corresponding amines.

$$CH_3—NH_2 \quad + \quad HCl \quad \longrightarrow \quad CH_3—NH_3^+ \quad + \quad Cl^-$$

Methylamine Methylammonium chloride
(amine salt)

D. Amides

When a carboxylic acid reacts with ammonia or an amine, the product is an amide. The functional group, called the amide group, and some examples of amides are shown below:

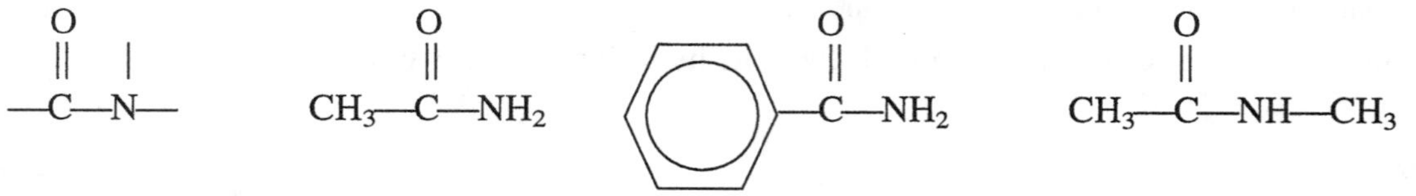

Amide   Acetamide   Benzamide   *N*-Methylacetamide

In a reaction called *amidation*, an amide forms when a carboxylic acid is heated with ammonia or an alkyl or aromatic amine.

$$CH_3—\overset{\displaystyle O}{\overset{\|}{C}}—OH \quad + \quad CH_3—NH_2 \quad \longrightarrow \quad CH_3—\overset{\displaystyle O}{\overset{\|}{C}}—NH—CH_3$$

Acetic acid   Methylamine   *N*-Methylacetamide

Hydrolysis of an Amide

When an amide is hydrolyzed, the amide bond is broken and the carboxylic acid and the amine are separated. Hydrolysis takes place in either an acid or a base. Acid hydrolysis produces the carboxylic acid and ammonium salt. In a base, the hydrolysis reaction produces the salt of the carboxylic acid and ammonia. The odor of ammonia and the reaction of ammonia with litmus paper are used to detect the hydrolysis reaction.

$$CH_3—\overset{\displaystyle O}{\overset{\|}{C}}—NH_2 \quad + \quad HCl \quad \longrightarrow \quad CH_3—\overset{\displaystyle O}{\overset{\|}{C}}—OH \quad + \quad NH_4Cl$$

Acetamide   Acetic acid   Ammonium chloride

$$CH_3—\overset{\displaystyle O}{\overset{\|}{C}}—NH_2 \quad + \quad NaOH \quad \longrightarrow \quad CH_3—\overset{\displaystyle O}{\overset{\|}{C}}—O^- Na^+ \quad + \quad NH_3(g)$$

Acetamide   Sodium acetate   Ammonia

Lab Information

Time: 2 ½–3 hr

Comments: Some amines have an irritating odor. Work in the hood. Tear out the report sheets and place them next to the matching procedures.

Related topics: Amines, solubility and pH of amines, amidation, amides, hydrolysis of amides

Experimental Procedures

WEAR YOUR PROTECTIVE GOGGLES!

A. Structure and Classification of Amines

Materials: Organic model kits

A.1 Use a model kit to prepare or observe models of ammonia, methylamine, dimethylamine, trimethylamine, and aniline. Write the condensed structural formulas of each model in the laboratory report.

A.2 Classify each amine as primary (1°), secondary (2°), or tertiary (3°).

B. Solubility of Amines in Water

Materials: Aniline, *N*-methylaniline, triethylamine, test tubes, test tube rack, stirring rod, pH paper

WORK IN THE HOOD. THE VAPORS OF AMINES ARE IRRITATING TO THE NOSE AND SINUSES.

B.1 To three separate test tubes, add 5 drops of aniline, *N*-methylaniline, and triethylamine. Draw the condensed structural formulas of each amine and state its classification (1°, 2°, 3°).

B.2 Cautiously note the odor of each. Remember to hold a fresh breath of air while you fan the vapor toward you. Record.

B.3 Add 2 mL of water to each test tube. Stir. Describe their solubility in water.

B.4 Determine the pH of each solution. Dip a stirring rod in each amine solution and then touch it to pH paper. Record. *Save these test tubes and samples for part C.*

C. Neutralization of Amines with Acids

Materials: Test tubes from part B, blue litmus paper, 10% HCl

C.1 Add 10% HCl dropwise to the amine solution until the solution is acidic to litmus paper. Record any changes in solubility of each amine. Note any changes in odor.

C.2 Write and balance equations for the neutralization of aniline, *N*-methylaniline, and triethylamine with HCl.

D. Amides

Materials: Organic model kit, acetamide, benzamide, test tubes, spatula, 10% HCl, 10% NaOH, 250-mL beaker for water bath, hot plate, red litmus paper

D.1 Make a model of acetic acid and ammonia. Show how the structures change in the formation of acetamide. Write the equation for the amidation.

D.2 Place small amounts (tip of a spatula) of acetamide and benzamide in separate test tubes. *Cautiously* note the odor. Add 2 mL of water to each. Record the solubility of each in water.

       Acetamide   Benzamide

Hydrolysis of an Amide

D.3 Make a model of acetamide and HCl. Use the models to show the hydrolysis reaction of the amide with HCl, and write the equation.

D.4 Using the test tubes from D.2, add 2 mL of 10% HCl to each. Place the test tubes in a boiling water bath and heat gently for 5 minutes. Cautiously note any odor coming from each mixture. Record your observations.

D.5 Prepare two more test tubes as you did in part D.2. Add 2 mL of 10% NaOH to each. Place the test tubes in a boiling water bath. Wet a piece of pH paper or red litmus paper and hold over the mouth of each test tube. Heat the tubes gently for 5 minutes. Record any change in the color of the litmus paper. Cautiously note any odor coming from each mixture. Record your observations. Write the equation for the hydrolysis of acetamide in base.

Report Sheet - Lab 33

Date _____ Name _____

Section _____ Team _____

Instructor _____

Pre-Lab Study Questions

1. What is the functional group in amines? In amides?

2. What products are formed when amides are hydrolyzed?

3. How is the amide bond important in proteins?

A. Structure and Classification of Amines

Compound	A.1 Condensed Structural Formula	A.2 Classification (1°, 2°, 3°)
Ammonia		✕
Methylamine		
Dimethylamine		
Trimethylamine		
Aniline		

Questions and Problems

Q.1 In the discussion, the structures are given for Neo-Synephrine and methamphetamine. Give the amine classification of each of the compounds.

Report Sheet - Lab 33

B. Solubility of Amines in Water

		Aniline	N-Methylaniline	Triethylamine
B.1	Condensed structural formula			
	Classification 1°, 2°, 3°			
B.2	Odor			
B.3	Solubility in water			
B.4	pH			

Questions and Problems

Q.2   What type of compound accounts for the "fishy" odor of fish?

Q.3   Explain why amines are basic.

Report Sheet - Lab 33

C. Neutralization of Amines with Acids

	Aniline	*N*-Methylaniline	Triethylamine
C.1 Solubility after adding HCl			
Odor after adding HCl			

C.2 Equation for the neutralization of aniline with HCl

Equation for the neutralization of *N*-methylaniline with HCl

Equation for the neutralization of triethylamine with HCl

Questions and Problems

Q.4 How does lemon juice remove the odor of fish?

Q.5 Write an equation for the reaction of butylamine with HCl.

Report Sheet - Lab 33

D. Amides

D.1 Equation for the formation of acetamide

	Acetamide	Benzamide
D.2 Odor		
Solubility		

D.3 Equation for the hydrolysis of acetamide in acid.

	Acetamide	Benzamide
D.4 Odor after adding HCl		
D.5 Odor after adding NaOH		
Change in red litmus paper		

Equation for the hydrolysis of acetamide in base.

Questions and Problems

Q.6 You have unknowns that are a carboxylic acide, an ester, and an amine. Describe how you would distinguish among them.

Goals

- Write equations for amidation and hydrolysis reactions.
- Prepare the common analgesic acetaminophen.
- Use solubility to purify a crude sample of acetanilide.

Discussion

A. Synthesis of Acetaminophen

Compounds used to relieve pain are called analgesics and compounds used to reduce a fever are antipyretics. Aspirin is both an analgesic and antipyretic and so is acetaminophen, which is an amide.

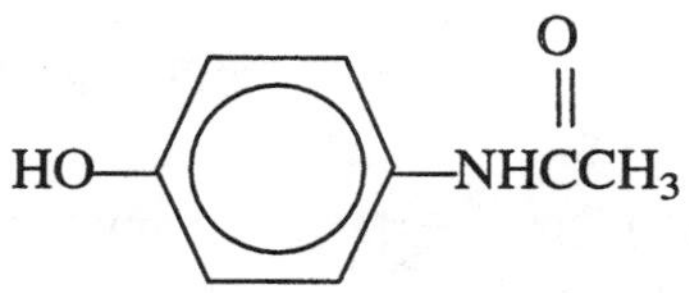

$$HO-\bigcirc-NHCCH_3$$
$$(C=O)$$

Acetaminophen

People who are sensitive to aspirin may use products such as Tylenol (acetaminophen) or Motrin, Advil, and Nuprin, which contain ibuprofen. Ibuprofen is a carboxylic acid, not an amide.

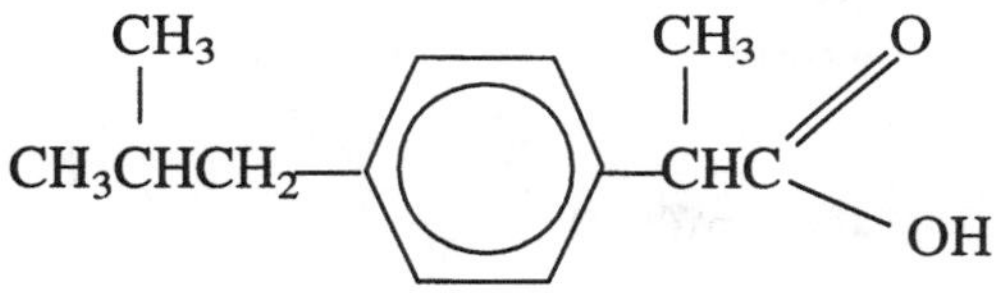

$$CH_3CHCH_2-\bigcirc-CHC \begin{smallmatrix} O \\ OH \end{smallmatrix}$$

Ibuprofen

Aspirin and acetaminophen are used in several common analgesic preparations, which may also contain caffeine and buffers. See Table 34.1.

Table 34.1 *Some Products with Aspirin and/or Acetaminophen*

Product	Aspirin	Acetaminophen	Caffeine
Aspirin	325 mg		
Anacin	400 mg		32 mg
Bufferin	324 mg		32 mg
Excedrin	250 mg	250 mg	65 mg
Tylenol		325 mg	
Side effect	Stomach upset; longer bleeding time; possilble toxic levels	Possible liver damage in high dosages for long-term users	Increase in pulse and heart rate

Marketed as Tylenol, acetaminophen is an amide that we can prepare in the laboratory from *p*-aminophenol and a two-carbon group obtained from acetic anhydride.

| *p*-Aminophenol | Acetic anhydride | Acetaminophen | Acetic acid |

Percent Yield

The maximum yield of acetaminophen from 1.5 g of *p*-aminophenol is calculated using the molar mass of the reactant *p*-aminophenol (109 g/mole) and the product acetaminophen (151 g/mole).

1.5 g *p*-aminophenol × $\dfrac{\text{1 mole } p\text{-aminophenol}}{\text{109 g } p\text{-aminophenol}}$ × $\dfrac{\text{1 mole acetaminophen}}{\text{1 mole } p\text{-aminophenol}}$ × $\dfrac{\text{151 g acetaminophen}}{\text{1 mole acetaminophen}}$

or 1.5 g *p*-aminophenol × $\dfrac{\text{151 g acetaminophen}}{\text{109 g } p\text{-aminophenol}}$ = g acetaminophen (possible)

The percent yield is calculated by dividing the actual mass of acetaminophen obtained by the maximum possible yield.

% Yield = $\dfrac{\text{mass of purified acetaminophen}}{\text{mass of acetaminophen possible}}$ × 100%

B.   Isolating Acetanilide from an Impure Sample

Acetanilide shows a tenfold increase in solubility from 25°C to 100°C. This difference in solubility can be used to isolate and purify acetanilide from an impure sample.

Acetanilide

The percent yield is calculated by dividing the mass of the purified acetanilide by the mass of the impure sample.

% Yield = $\dfrac{\text{mass of purified product}}{\text{mass of impure sample}}$ × 100%

Lab Information

Time: 2 ½–3 hr

Comments: When recrystallizing product, use ice water to rinse crystals.
                Tear out the report sheets and place them beside the matching procedures.

Related topics: Amides, amidation

Experimental Procedures

 **GOGGLES REQUIRED!**

A. Synthesis of Acetaminophen

Materials: 125-mL Erlenmeyer flask, *p*-aminophenol, 85% H_3PO_4, acetic anhydride, 150-mL beaker, hot plate, stirring rod, ice bath, Büchner filtration apparatus, filter paper, melting point apparatus, watch glass

A.1 Weigh a 125-mL Erlenmeyer flask. Add 1.5 g of *p*-aminophenol and reweigh. *Avoid contact with skin. You may wish to wear gloves.* Add 25 mL of water and 20 drops of 85% H_3PO_4. Then heat the flask on a hot plate to boiling. The *p*-aminophenol should dissolve. Remove the flask from the hot plate. ***Working in the hood, carefully*** add 2 mL of acetic anhydride. Stir. Place the flask in an ice bath. Stir to crystallize the acetaminophen. You may need to scratch the walls of the flask to start the crystallization. If no crystals appear, add a small crystal of acetaminophen to start the formation of solid acetaminophen. Allow the flask to stay in the ice-water bath for 30 minutes.

A.2 Collect the crystals using a Büchner funnel vacuum. (See Lab, "Synthesis of Aspirin," for directions on using the Büchner filtration setup.) Wash the product with 10 mL of ice water. Allow the crystals to dry. Weigh a watch glass. Transfer the crude product to the watch glass and reweigh.

A.3 The melting point (mp) of acetaminophen is 169–171°C; p-aminophenol melts at 189–190°C. If a melting point apparatus is available, determine the melting point of your dry, crude acetaminophen product. Compare your melting point results with the known melting points for the starting and expected products.

Purification of Crude Acetaminophen (Optional)

A.4 Place the crude acetaminophen in a 150-mL beaker. Add 20 mL of water and heat on a hot plate until all of the solid dissolves. If the solution reaches boiling and crystals remain, add more water, a few mL at a time. Remove the flask and allow the solution to cool. When crystals begin to appear, place the flask in an ice bath for 20 minutes. If no crystals appear, scratch the inside walls. Collect the crystals using the Büchner filtration apparatus. Wash with 10 mL of cold water. Transfer the filter paper and crystals to a paper towel and let dry. Weigh a watch glass. Transfer the pure product to the watch glass and reweigh.

A.5 Calculate the percentage yield of the pure acetaminophen.

A.6 Determine the melting point of the recrystallized acetaminophen. How does the mp of the pure product compare with the mp of the crude product?

B. Isolating Acetanilide from an Impure Sample

Materials: Two 250-mL beakers, impure acetanilide, hot plate, short-stem funnel, filter
paper, iron ring, Büchner funnel, watch glass, small vial and stopper

B.1 Weigh a 250-mL beaker. Add about 2 g of impure acetanilide. Weigh the beaker and contents.
Record. Calculate the mass of the impure sample. Add 50 mL of water to the beaker and heat
the mixture on a hot plate until no more solid material appears to dissolve.

While the mixture is heating, heat another beaker of water for use in the filtration. Place a
short-stem funnel fitted with filter paper in an iron ring. When you are ready to filter your
impure sample, pour some of the hot water through the funnel to warm the glass. Discard the
water.

Filter the warm sample of impure acetanilide into a clean beaker or flask. Rinse the funnel
with hot water so that crystals do not form in it. Place the beaker or flask containing the
warm filtrate in an ice bath. As the filtrate cools, crystals of acetanilide should form.

B.2 Weigh a watch glass. Collect the crystals of acetanilide using the Büchner funnel and suction
filtration apparatus. Place the crystals on the weighed watch glass and let them dry. Weigh
the watch glass with the dry product. Record the appearance of the purified acetanilide.

B.3 Determine the mass of the acetanilide and calculate the percent yield of the pure product.

B.4 Use a melting point apparatus to determine the melting point of the purified acetanilide.
Record.

Optional: Place the purified product in a small vial, stopper, and label with your name, %
yield, and melting point of the product. Turn in the product to your instructor.

Report Sheet - Lab 34

Date _____   Name _____

Section _____   Team _____

Instructor _____

Pre-Lab Study Questions

1. Draw the structures of acetaminophen and acetanilide. Describe their similarities.

2. What method is used to remove the impurities in a product?

A. Synthesis of Acetaminophen

A.1 Mass of flask   _____

Mass of flask and *p*-aminophenol   _____

A.2 Mass of watch glass   _____

Mass of watch glass and acetaminophen product   _____

A.3. Melting point of crude acetaminophen   _____°C

A.4 Mass of watch glass + pure product   _____

Mass of watch glass   _____

Mass of pure product   _____

A.5 Percent yield   _____%
Show calculations.

A.6 Melting point of purified acetaminophen product _____°C

Handbook value for melting point of acetaminophen   _____°C

Questions and Problems

Q.1 Acetaminophen does not act as a base in water, but p-aminophenol does. Explain.

Report Sheet - Lab 34

Questions and Problems

Q.2 What does the comparison of the melting points of your product(s) and the known melting point of acetaminophen tell you about the purity of your product?

Q.3 Phenacetin is another over-the-counter medication for reducing fever and relieving pain.

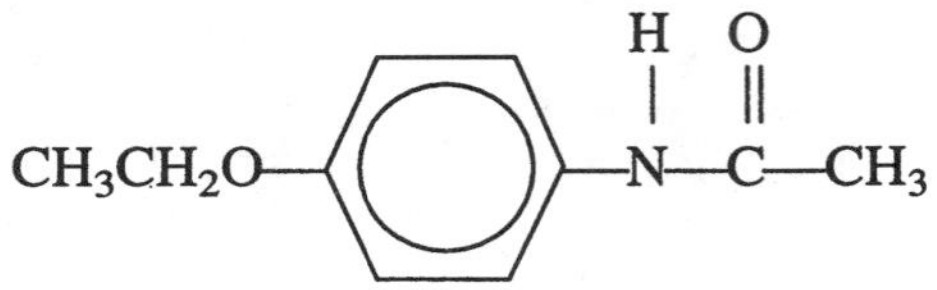

$$CH_3CH_2O-\text{⟨○⟩}-\overset{H}{\underset{|}{N}}-\overset{O}{\underset{||}{C}}-CH_3$$

Phenacetin

a. What are the functional groups in phenacetin?

b. How might you prepare phenacetin?

B. Isolating Acetanilide from an Impure Sample

B.1 Mass of beaker _____

Mass of beaker and impure acetanilide _____

B.2 Mass of watch glass _____

Mass of watch glass and acetanilide product _____

Appearance of purified acetanilide _____

B.3 Mass of purified acetanilide _____

Percent yield of the pure product _____ %
Show calculations.

B.4 Melting point of acetanilide product _____ °C

Handbook value for melting point of acetanilide _____ °C

Plastics and Polymerization

Goals

- Identify the monomer units in a polymer.
- Write a portion of a polymer from its monomer units.
- Identify the type of polymer in a plastic item from its recycling code.
- Prepare polymers of nylon, polystyrene, and Slime.

Discussion

Polymers are huge molecules that are made by combining many small molecules called monomers. Polymers are prevalent in nature. Cellulose and starch are polymers of glucose, a monosaccharide, and silk and wool as well as the enzymes in our cells, are proteins polymers composed of amino acids. In the last century, scientists have developed many kinds of *synthetic* polymers that are important in our daily lives. Some include the nonstick coating on cooking sheets and pans, foam rubber, disposable diapers, plastic cups, garden hoses, nylon, outdoor clothing and carpeting, plastic wrap, computer disks, and surfboards.

Polyethylene is a common polymer used to make plastic bottles, film, and plastic dinnerware. During polymerization, the double bonds of ethene (ethylene) molecules open and add to the next monomer. When large numbers of monomers combine to form a long carbon chain, it is a polymer of polyethylene.

ethene (ethylene) monomers → polyethylene section

A. Classification of Plastics

Because many of the plastic synthetic polymers are based on alkanes, the polymers do not decompose easily. They are not biodegradable and contribute to pollution. Recycling programs collect plastic materials and recycle the materials rather than adding them to landfills. Plastic items are labeled on the bottom with a recycling code so they can be sorted according to the type of plastic. (See Table 35.1.)

Table 35.1 *Types of Plastics and Their Recycling Codes*

Recycling Codes	Polymer	Examples
1 PETE	Polyethylene terephthalate	Soda bottles, carpets
2 HDPE	High density polyethylene	Milk and detergent bottles
3 V	Vinyl/Polyvinyl chloride	Plastic pipes, garbage bags
4 LDPE	Low density polyethylene	Soft bottles, carpets, dry cleaners bags
5 PP	Polypropylene	Raincoats, yogurt containers, artificial joints
6 PS	Polystyrene	Coffee cups, foam cartons, toys
7 Other	Resins, mixed polymers	Ketchup bottles

In this lab, we will identify an assortment of plastic items by their recycling codes and look at some of their physical properties. The different types of polymers vary in properties such as density. In recycling plants, plastics are shredded and added to a liquid such as water. Plastics denser than water sink, and those less dense float. Those that sink are recovered separately from those that float.

B. Gluep and Slime®

Slime® is a gel that is popular polymer product enjoyed by children. In the laboratory it is made from polyvinyl alcohol and a saturated borax ($Na_2B_4O_7$) solution. A similar type of gel can be made from Elmer's glue, which contains polyvinyl acetate rather than polyvinyl alcohol. Polyvinyl acetate is a polymer of vinyl acetate.

Vinyl alcohol → polymerization → Polyvinyl alcohol

Vinyl acetate → polymerization → Polyvinyl acetate

When borate from Borax is added to polyvinyl alcohol or polyvinyl acetate, large numbers of cross links form a viscous substance. The number of cross links, which are hydrogen bonds, can be increased or decreased by changing the borate concentration. They also break and reform with handling or from the weight of the gel. Thus the gel tends to flow, change shape, and break if it is pulled apart suddenly. Adding a few drops of acid destroys the gel; adding some NaOH reforms the gel.

C. Polystyrene

Styrene is the compound that is polymerized to give polystyrene.

Styrene → Polystyrene section

Polystyrene is a clear, brittle plastic used to make plastic glasses and coffee cups. In the polystyrene polymer, there may be as many as 3000 styrene monomers. At high temperatures, the polymerization of styrene is spontaneous, but at lower temperature an initiator such as benzoyl peroxide, which forms a high-energy radical, is required. You may be familiar with the compound benzoyl peroxide, which is used in many skincare creams for treatment of acne.

To prevent spontaneous polymerization of styrene in the bottle on a shelf, the manufacturer adds a small amount of an inhibitor, which is 4-*tert*-butylcatechol. This inhibitor is removed by passing styrene through some alumina, which absorbs the inhibitor.

Benzoyl peroxide → Benzoyloxy radial initiator

D.   Nylon

Nylon was introduced in 1938, which makes it one of the first synthetic polymers. Polymers known as nylon are polyamides made by the condensation of diamines and dicarboxylic acids. Nylon 6,6 is the condensation of hexamethylene diamine and adipic acid and nylon 6,10 uses hexamethylene diamine and sebacic acid. Two numbers indicating the number of carbon atoms designate the different nylons, the first for the diamine and the second for the dicarboxylic acid. Thus in nylon 6,10 the diamine has 6 carbons, and the dicarboxylic acid has 10 carbons

In this experiment, sebacoyl chloride rather than sebacic acid is heated with the amine. The two reactants will be present in two layers. The polymer will form a film at the interface where the diamine and the diacid chloride are in contact.

325

Lab Information

Time: 2-3 hr
Comments: Tear out the Lab report sheets and place them beside the matching procedures.
Related topics: monomers, polymers, and polymerization

Experimental Procedures

WEAR YOUR PROTECTIVE GOGGLES!

A. Classification of Plastics

Materials: samples of plastic items such as Styrofoam cups, Saran wrap, milk cartons, nylon, transparencies, garden hose, plastic glasses (smash with a hammer), yogurt containers, bubble wrap, an alcohol-water mixture (9 mL 70% isopropyl alcohol (rubbing alcohol) and 6 mL water), stirring rods, forceps, acetone, two 250-mL beakers, test tubes, and test tube rack, small pieces of each type of plastic, unknowns (optional), Bunsen burner

A.1 List the type of item and its use. Record its recycling code. Write the name of the type of plastic. Using a reference such as your textbook, write the name of the monomer for the plastic.

A.2 *Density Test*
Fill a 250-mL beaker about $\frac{1}{2}$ full with water. Fill a second 250-mL beaker about $\frac{1}{2}$ full with an isopropyl alcohol:water (3:2) mixture. Obtain 2 small pieces of each type of plastic. Add one piece to each beaker. If the plastic piece floats, use a stirring rod to gently push the plastic piece below the surface. Record whether each type of plastic is more dense or less dense than water (1.0 g/mL) and more dense or less dense than water-alcohol mixture.

Obtain an unknown piece of plastic. Describe its properties. Determine its results in the density test. Identify the types of plastic it may be.

A.3 *Solubility In Acetone Test*   ***This may be a demonstration set up in the hood.***
Caution: Acetone is flammable. Work in the hood. Obtain a sample of each type of plastic. Place 4 mL of acetone in each of 6 test tubes in a test tube rack. Add a piece of one type of plastic to each test tube. Allow the plastic piece to remain in the acetone for 10 minutes. Use forceps to remove the plastic piece and rinse with water. Press between your fingers. Some plastics "swell" in organic solvents and become soft and pliable. Compare the plastic pieces before and after acetone treatment. Record any changes in the appearance of each type of plastic.

Test another piece of the same unknown plastic as in A.2. Identify the types of plastic it may be. Discard in a waste container. Discard the acetone in a waste acetone container.

A.4 *Combustion Test*   **Work in the hood**
Obtain small pieces of each type of plastic. Using forceps, hold a piece in the flame of a Bunsen burner to determine if the plastic ignites. If it begins to burn, remove it from the flame and allow it to continue to burn. Observe how fast it burns and the color of its flame and residue.
Test another piece of your same unknown plastic. Identify the types of plastic it may be. Place the plastic residue in a waste container.

B. Gluep and Slime®

Materials: 10-mL and 50-mL graduated cylinders, Styrofoam cup, plastic sticks or spatulas, saturated borate solution, Elmer's glue, lab gloves, 4 % polyvinyl alcohol solution

Gluep

B.1 Obtain a Styrofoam cup. Place 20 mL of Elmer's glue and 20 ml of water in the cup. Sir to mix. Add 10 of the borate solution to the cup while stirring vigorously. Continue to stir for at least 10 minutes until it is a viscous gel. Wearing gloves, remove the gel and knead it like bread for 3-4 minutes.

Observe the physical properties of the gel. What does it smell like? What is its texture? What happens when you stretch it slowly? Fast? What happens if you roll it into a ball and let it sit on a piece of paper on the lab bench? Record your observations. What happens if you make a ball of the gel and drop it onto the lab bench? What happens if you roll the gel into a long, thin roll and pull the ends apart?

The following may be done by three lab teams each choosing one procedure.

B.2 Repeat procedure B.1 but use 40 mL of water.

B.3 Repeat procedure B.1 but do not add water.

B.4 Repeat procedure B.1 but use 20 mL of borate solution.

Slime®

B.5 Obtain a Styrofoam cup. Add 20 mL of 4% polyvinyl alcohol solution and 5 mL of the borate solution. Stir for 5 minutes. Remove and knead like bread. Compare the properties of Slime to the Gluep product.

C. Polystyrene

Materials: Styrene, benzoyl peroxide (or an acne preparation which contains 5% or 10% benzoyl peroxide), stirring rods, alumina, funnel, filter paper, hot plate, heavy-duty aluminum foil, 10- or 20-mL beaker, 50-mL beaker, wood stick

1. Place 0.3-0.4 g of alumina in a test tube. Carefully add 4 mL of styrene, which contains inhibitor.

2. Fold a filter paper. Pour the styrene-alumina mixture into the filter paper and collect the styrene (inhibitor removed) in a small beaker. Place used alumina in a nonhazardous waste container.

3. Obtain two square sheets of aluminum foil 10 cm × 10 cm. Using a double layer of the aluminum sheets, form a mold by molding them around the bottom of a small beaker. Place this mold in a 50-mL beaker.

4. Prepare a boiling water bath using a hot plate.

5. Weigh out 0.050 g of benzoyl peroxide (or 0.2 g of a 10% benzoyl peroxide acne cream preparation) and add to the styrene (filtered). Swirl the beaker 3-4 minutes. Pour the mixture into the aluminum foil mold in the 50-mL beaker. Clamp the beaker containing the foil mold and styrene mixture to the ring stand. Lower the bottom of the beaker into the boiling water bath. Be careful not to spill the styrene mixture out of the foil mold as you set this up.

Leave the beaker in the water bath for about 45 minutes. Add more water to the water bath as need-ed. Occasionally stir the styrene with a wooden stick. As the temperature rises, the mixture should become more viscous. *During this heating time, proceed to the next lab, which is the synthesis of nylon.*

6. Remove the beaker and the aluminum mold with the styrene product from the water bath, and allow it to cool slowly to room temperature. Then place the beaker in an ice bath for 20 minutes. If you leave the wooden stick in the styrene, the styrene will solidify around it. Remove the foil mold from the beaker, and separate the aluminum foil from around the polystyrene. You may need to let the polymer harden until the next laboratory time.

C.1   Describe the appearance and texture of the polystyrene product.

D. Nylon

This experiment may be done in pairs of lab teams.

Caution: Use a fume hood. The reactants are irritating to the skin and eyes.

Hexamethylenediamine and sebacoyl chloride are irritating to the skin, eyes, and respiratory sys-tem. Sodium hydroxide is extremely caustic and can cause severe burns. Contact with the skin and eyes must be prevented. Hexane is extremely flammable. Hexane vapor can irritate the respiratory tract and, in high concentrations, be narcotic.

Materials: 50-mL, 100-mL, and 250 beakers, 10-mL and 50-mL graduated cylinders, forceps, metal spatula, large test tube (for spooling the nylon), stirring rods, gloves (must not dissolve in hexane), 50% aqueous ethanol solution, 6 M HCl, 6 M NaOH

Solution 1: 4% hexamethylenediamine and NaOH. If solid, place the reagent bottle in hot water to melt (mp 39°C).

Solution 2: 4% sebacoyl chloride, $ClCO(CH_2)_8COCl$, in hexane

1. Obtain a clean 50-mL beaker. Add 20 mL of solution 1 (1,6-hexamethylenediamine and NaOH). Using a 50-mL graduated cylinder, obtain 20 mL of solution 2 (4% sebacoyl chloride in hexane).

> ### *Put on gloves. Work in the hood*

2. Tilt the beaker containing the hexamethylenediamine solution and slowly pour the sebacoyl chloride solution down the wall of the beaker. Allow the mixture to site undisturbed for 1 minute. Two layers will form with a whitish film of nylon polymer film at the interface. Run a metal spatula around the walls of the beaker to loosen the film from the beaker sides.

3. Carefully lower the tip of a pair of forceps through the top layer to take hold of the center of the nylon film. Slowly pull a strand of the polymer upward out of the solution. A continuous section of nylon should form. Secure the end of the strand in the forceps by wrapping it around the center of the large test tube. Holding the test tube, turn it slowly to draw out a continuous nylon strand from the interface. You should be able to draw out most of the interface film. If the strand breaks, use forceps to start a new one. Rinse the nylon strand on the test tube under tap water to remove excess reactants. *Do not touch the nylon strand before you have thoroughly washed it with water.*

4. Use a metal spatula to detach the nylon loop from the test tube and slide the nylon loop into a beaker containing 20 mL of 50% aqueous alcohol solution to remove any remaining reactants. Rinse with water again. Using forceps grab an end of the nylon and stretch out the nylon strand and place it on paper towels until it is dry.

D.1   Describe the appearance and texture of the nylon polymer.

D.2   Cut a 15-20 cm section of nylon and pull on both ends. When you stretch the nylon, does it break or return to its original form (elastic)? Record your observations.

D.3   Cut several 2-3 cm sections of nylon. Place one piece in a test tube containing 5 mL of 6M HCl. Place a second piece in a test tube containing 5 mL of 6M NaOH. Place a third piece in a test tube containing 5 mL of acetone. Observe any changes in the nylon in the presence of a strong acid, a strong base, and an organic solvent. Record your observations.

D.4   Light a Bunsen burner and use forceps to hold the nylon piece in the flame. Record your observations on the flammability of nylon.

Disposal

Stir the remaining reaction mixture to form a ball of nylon and discard in the proper solid waste container. Use a metal spatula to remove any remaining nylon polymer from the glass. Place remaining solvents in the proper waste liquid container. Dispose of acid and base solution by pouring them into the sink and rinsing with plenty of water. Dispose of acetone in the proper waste container.

Report Sheet - Lab 35

Date _____   Name _____

Section _____   Team _____

Instructor _____

Pre-Lab Study Questions

1. Why do most plastics remain in landfills for a long time?

2. What is the purpose of the recycling codes on plastics?

A. Classification of Plastics

A.1 Plastic Item/Use	Recycling code	Name of plastic	Monomer unit (name)

Type of Plastic	A.2 Water	A.2 Water-Alcohol	A.3 Acetone	A.4 Combustion
PETE				
HDPE				
V				
LDPE				
PP				
PS				
Unknown				
Possible plasic types				

What type(s) of plastic is your unknown? _____

Report Sheet - Lab 35

Questions and Problems

Q.1 If you are going to make life preservers, which types of plastic could you use? Why?

Q.2 Which type(s) of plastic would you not use for storing left over acetone? Why?

Q.3 You are going into a small business to manufacture handles for a barbecue. Which types of plastic could you use?

B. Gluep and Slime®

Properties	Gluep Samples				Slime®
	B.1	B.2	B.3	B.4	B.5
Smell and texture					
Stretching slowly and then fast					
Letting it sit on the lab bench					
Dropping it on the lab bench					
Pulling it apart					
Other					

Report Sheet - Lab 35

Questions and Problems

Q.4 What combination of reactants for Gluep was the best to work with? Why?

Q.5 What combination of reactants for Gluep did not work well?

Q.6 How does the "slime" product compare to your Gluep product?

C. Polystyrene

C.1 Appearance and texture of your polystyrene sample.

D. Nylon

D.1 Appearance and texture of the nylon polymer.

D.2 Behavior of nylon when stretched.

D.3 Behavior of nylon in acid, base, and organic solvent

HCl	NaOH	Acetone

D.4 Describe the flammability of nylon.

Goals

- Use R groups to determine if an amino acid will be acidic, basic, or neutral; hydrophobic or hydrophilic.
- Use paper chromatography to separate and identify amino acids.
- Calculate R_f values for amino acids.

Discussion

A. Amino Acids

In our body, amino acids are used to build tissues, enzymes, skin, and hair. About half of the naturally occurring amino acids, the *essential amino acids,* must be obtained from the proteins in the diet because the body cannot synthesize them. Amino acids are similar in structure because each has an amino group (–NH$_2$) and a carboxylic acid group (–COOH). Individual amino acids have different organic groups *(R groups)* attached to the alpha carbon atom. Variations in the R groups determine whether an amino acid is hydrophilic or hydrophobic, and acidic, basic, or neutral.

R group

Amino group H—N—C—C Carboxylic **acid** group

Some R groups contain carbon and hydrogen atoms only, which makes the amino acids nonpolar and hydrophobic ("water-fearing"). Other R groups contain OH or SH atoms and provide a polar area that makes the amino acids soluble in water; they are hydrophilic ("water-loving"). Other hydrophilic amino acids contain R groups that are carboxylic acids (acidic) or amino groups (basic). The R groups of some amino acids used in this experiment are given in Table 36.1.

Table 36.1   *Amino Acids Found in Nature*

R Group	Amino Acid	Symbol	Polarity	Reaction to Water
H—	Glycine	Gly	Nonpolar	Hydrophobic
CH$_3$—	Alanine	Ala	Nonpolar	Hydrophobic
⬡—CH$_2$—	Phenylalanine	Phe	Nonpolar	Hydrophobic
HO—CH$_2$—	Serine	Ser	Polar	Hydrophilic
HOC—CH$_2$—	Aspartic acid	Asp	Polar, acidic	Hydrophilic
HOC—CH$_2$—CH$_2$—	Glutamic acid	Glu	Polar, acidic	Hydrophilic
H$_2$NCH$_2$CH$_2$CH$_2$CH$_2$—Lysine		Lys	Polar, basic	Hydrophilic

Ionization of Amino Acids

An amino acid can ionize when the carboxyl group donates a proton, and when the lone pair of electrons on the amino group attracts a proton. Then the carboxyl group has a negative charge, and the amino group has a positive charge. The ionized form of an amino acid, called a *zwitterion* or *dipolar ion,* has a net charge of zero.

$$
\underset{\text{Alanine}}{NH_2-\overset{\overset{\displaystyle CH_3}{|}}{CH}-\overset{\overset{\displaystyle O}{\|}}{C}-OH}
\qquad\qquad
\underset{\text{Zwitterion of alanine}}{\overset{+}{NH_3}-\overset{\overset{\displaystyle CH_3}{|}}{CH}-\overset{\overset{\displaystyle O}{\|}}{C}-O^-}
$$

In acidic solutions (low pH), the zwitterion *accepts* a proton (H^+) to form an ion with a positive charge. When placed in a basic solution (high pH), the zwitterion *donates* a proton (H^+) to form an ion with a negative charge. This is illustrated using alanine.

$$
\underset{\substack{\text{High pH}\\(\text{charge}=1-)}}{NH_2-\overset{\overset{\displaystyle CH_3}{|}}{CH}-\overset{\overset{\displaystyle O}{\|}}{C}-O^-}
\;\underset{\substack{\text{Base}\\ \text{Donates H+}}}{\longleftarrow}\;
\underset{\substack{\text{Zwitterion (neutral pH)}\\(\text{charge}=0)}}{\overset{+}{NH_3}-\overset{\overset{\displaystyle CH_3}{|}}{CH}-\overset{\overset{\displaystyle O}{\|}}{C}-O^-}
\;\underset{\substack{\text{Acid}\\ \text{accepts H+}}}{\longrightarrow}\;
\underset{\substack{\text{Low pH}\\(\text{charge}=1+)}}{\overset{+}{NH_3}-\overset{\overset{\displaystyle CH_3}{|}}{CH}-\overset{\overset{\displaystyle O}{\|}}{C}-OH}
$$

B.   Chromatography of Amino Acids

Chromatography is used to separate and identify the amino acids in a mixture. Small amounts of amino acids and unknowns are placed along one edge of Whatman #1 paper. The paper is then placed in a container with solvent. With the paper acting like a wick, the solvent flows up the chromatogram, carrying amino acids with it. Amino acids that are more soluble in the solvent will move higher on the paper. Those amino acids that are more attracted to the paper will remain closer to the origin line. After removing and drying the paper, the amino acids can be detected (visualized) by spraying the dried chromatogram with ninhydrin.

The distance each amino acid travels up the paper from the origin (starting line) is measured and the R_f values calculated. The R_f value is the distance traveled by an amino acid compared to the distance traveled by the solvent. See Figure 36.1.

$$R_f = \frac{\text{distance traveled by an amino acid (cm)}}{\text{distance traveled by the solvent (cm)}}$$

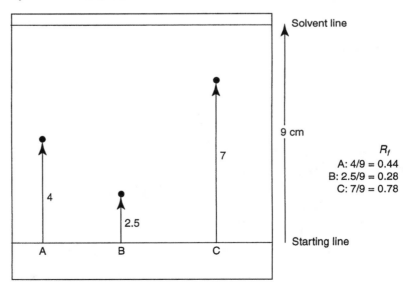

Figure 36.1 A developed chromatogram (R_f values calculated for A, B, and C)

To identify an unknown amino acid, its R_f value and color with ninhydrin is compared to the R_f values and colors of known amino acids in that solvent system. In this way, the amino acids present in an unknown mixture of amino acids can be separated and identified.

Lab Information

Time: 2 ¹/₂ –3 hr
Comments: Ninhydrin spray causes stains. Use it carefully.
                       Tear out the report sheets and place them next to the matching procedures.
Related Topics: Amino acids, zwitterions

Experimental Procedures

GOGGLES REQUIRED!

A. Amino Acids

Materials: Organic model kits or prepared models

A.1 Using an organic model kit, construct models of glycine and alanine. Draw their structures. Convert the alanine model to a model of serine. Indicate whether each of the amino acids would be hydrophobic or hydrophilic.

A.2 Form ionized (zwitterion) glycine by removing a H atom from the –COOH group and placing it on the N atom in the –NH₂ group. Draw the structure of the glycine zwitterion.

A.3 Write the structure of glycine in a base and in an acid.

B. Chromatography of Amino Acids

Materials: 600-mL beaker, plastic wrap, plastic gloves, Whatman chromatography paper #1 (12 cm × 24 cm), toothpicks or capillary tubing, drying oven (80°C) or hair dryer, metric ruler, stapler, amino acids (1% solutions): phenylalanine, alanine, glutamic acid, serine, lysine, aspartic acid, and unknown
Chromatography solvent: isopropyl alcohol, 0.5 M NH₄OH; 0.2% ninhydrin spray

Preparation of paper chromatogram Using forceps or plastic gloves (or a sandwich bag), pick up a piece of Whatman #1 chromatography paper that has been cut to a size of 12 cm × 24 cm. *Keep your fingers off the paper because amino acids can be transferred from the skin.* When this paper is rolled into a cylinder, it should fit into the chromatography tank (large beaker) without touching the sides. Draw a pencil (lead) line about 2 cm from the long edge of the paper. This will be the starting or origin line. Mark off seven points about 3 cm apart along the line. (See Figure 36.2.) Place your name or initials in the upper corner with the pencil.

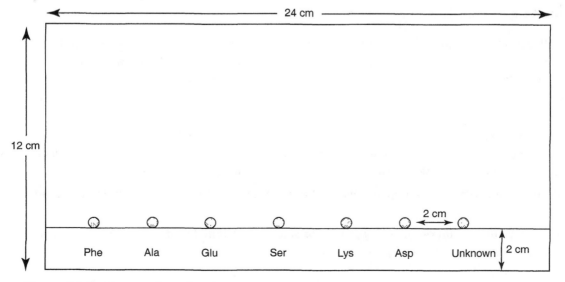

Figure 36.2 Preparation of a chromatogram

Application of amino acids Apply small amounts of the following 1% amino acid solutions: phenylalanine, alanine, glutamic acid, serine, lysine, and aspartic acid. Also prepare a spot of an unknown. Use the toothpick applicators or capillary tubes provided in each amino acid solution to make a small spot (the size of the letter **o**) by lightly touching the tip to the paper. After the spot dries, retouch the spot one or two more times to apply more amino acid, but keep the diameter of the spot as small as possible. A hair dryer can be used to dry the spots. *Always return the applicator to the same amino acid solution.* Using a pencil, label each spot as you go along. Allow the spots to dry.

Preparation of chromatography tank *Work in the hood.* Prepare the solvent by mixing 10 mL of 0.5 M NH_4OH and 20 mL of isopropyl alcohol. Pour the solvent into a 600-mL beaker to a depth of about 1 cm but not over 1.5 cm. (The height of the solvent must not exceed the height of the origin line on your chromatography paper.) Cover the beaker tightly with plastic wrap. This is your chromatography tank. Label the beaker with your name and leave it in the hood.

Running the chromatogram Roll the paper into a cylinder and staple the edges *without overlapping. The edges should not touch.* Slowly lower the cylinder into the solvent of the chromatography tank with the row of amino acids at the bottom. Make sure that the paper does not touch the sides of the beaker. See Figure 36.3. Cover the beaker with the plastic wrap and leave it undisturbed. The tank must not be disturbed while solvent flows up the paper. Let the solvent rise until it is 2–3 cm from the top edge of the paper. It may take 45–60 minutes. *Do not let the solvent run over the top of the paper.*

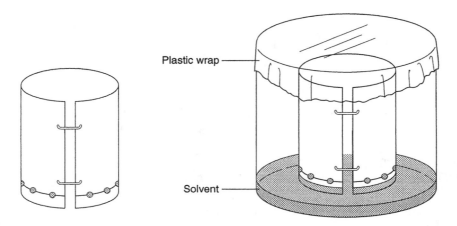

Plastic wrap

Solvent

Figure 36.3 Chromatogram in a solvent tank

Visualization of amino acids *Working in the hood,* carefully remove the paper from the tank. Take out the staples and spread the chromatogram out on a paper towel. *Immediately* mark the solvent line with a pencil. Allow the chromatogram to dry completely. A hair dryer or an oven of about 80°C may be used to speed up the drying process. Pour the solvent into a waste solvent container.

Working in the hood, spray the paper lightly, but evenly, with a ninhydrin solution. Dry the sprayed paper by placing it in a drying oven at about 80°C for 3–5 minutes or use a hair dryer. Distinct, colored spots will appear where the ninhydrin reacted with the amino acids.

Caution: Use the ninhydrin spray inside the hood. Do not breathe the fumes or get spray on your skin.

B.1 Draw the chromatogram on the report sheet, or staple the original to the report sheet. Record the color of each spot on the drawing or original.

B.2 Measure the distance (cm) from the starting line to the top of the solvent line to obtain the distance traveled by the solvent.

B.3 Outline each spot with a pencil. Place a dot at the center of each spot. Measure the distance in centimeters (cm) from the origin to the center dot of each spot.

B.4 Calculate and record the R_f values for the known amino acid samples and the unknown amino acid.

$$R_f = \frac{\text{distance traveled by an amino acid}}{\text{distance traveled by the solvent}}$$

B.5 **Identification of unknown amino acids** Compare the color and R_f values produced by the unknown amino acids. Identical amino acids will give similar R_f values and form the same color with ninhydrin. Identify the amino acid(s) in the unknown.

Report Sheet - Lab 36

Date _____   Name _____

Section _____   Team _____

Instructor _____

Pre-Lab Study Questions

1. What are the functional groups in all amino acids?

2. How does an R group determine if an amino acid is acidic, basic, or nonpolar?

A. Amino Acids

A.1	Structures of Amino Acids	
Glycine	**Alanine**	**Serine**
Hydrophobic or hydrophilic?		

A.2 Zwitterion structure of glycine	A.3 Glycine ion in base	Glycine ion in acid

Report Sheet - Lab 36

Questions and Problems

Q.1 Write the structure of the zwitterion of alanine.

Q.2 Write the prevalent form of alanine in an acidic solution.

B. Chromatography of Amino Acids

B.1 Chromatogram drawing or original, with colors of spots written in

| |
| |
| |
| |
| |
| |

Calculation of R_f Values

B.2 Distance from origin to solvent line: _____

Amino acid	Color	B.3 **Distance (cm)** amino acid traveled	B.4 R_f **value**
Phenylalanine			
Alanine			
Glutamic acid			
Serine			
Lysine			
Aspartic acid			
Unknown			

B.5 Identification of unknown # _____ : _____

Peptides and Proteins

Goals

- Identify the structural patterns of proteins.
- Observe the denaturation of proteins.
- Use the isoelectric point of casein in milk to isolate the protein.
- Use chemical tests to identify proteins and amino acids.

Discussion

A. Peptide Bonds

A dipeptide forms when two amino acids bond together. A peptide (or amide) bond forms between the carboxylic acid of one amino acid and the amino group of the next amino acid with the loss of H_2O.

Amino acid (1) Amino acid (2) Dipeptide

In the reverse reaction, hydrolysis, water adds to the peptide bond to yield the individual amino acids. For example, the dipeptide glycylalanine hydrolyzes to give the amino acids glycine and alanine.

Glycylalanine (Gly-Ala), a dipeptide   Glycine   Alanine

B. Structure of Proteins

When many amino acids are joined by peptide bonds they make a polypeptide. If more than 50 amino acids are in the peptide chain, it is usually considered to be a protein. Proteins make up many important features in the body including skin, muscle, cartilage, hair, fingernails, enzymes, and hormones. Proteins have specific structures, which are determined by the sequence of the amino acids. The peptide bonds that join one amino acid to the next are the *primary* level of protein structure. *Secondary* structures include the alpha (α) helix formed by the coiling of the peptide chain, or a β-pleated sheet structure formed between protein strands.

The secondary structures are held in place by many hydrogen bonds between the oxygen atoms of the carbonyl groups and the hydrogen atoms of the amide group. At the *tertiary* level, interactions between the side groups such as ionic bonds or salt bridges, disulfide bonds, and hydrophilic bonds give the protein a compact shape. Such tertiary structures are evident in the spherical shape of globular proteins. Similar interactions between two or more tertiary units produce the *quaternary* structure of many active proteins.

C. Denaturation of Proteins

Denaturation of a protein occurs when certain conditions or agents disrupt the bonds that hold together the secondary or tertiary structures of a protein. Proteins are denatured with heat, acid, base, ethanol, tannic acid, and heavy metal ions (silver, lead, and mercury). In most cases, the protein coagulates. Heat increases the motion of the atoms and disrupts the hydrogen bonds and the hydrophobic (nonpolar) attractions. Strong acids and bases disrupt the ionic bonds between amino acids with acidic and basic side groups. Alcohol, an organic solvent, destroys hydrogen bonds. The heavy metal ions, Ag^+, Pb^{2+}, and Hg^{2+}, react with the sulfur groups and carboxylic acid groups, which prevents the formation of the cross-links for tertiary structures and quaternary structures.

D. Isolation of Casein (Milk Protein)

A typical source of protein is milk, which contains the protein casein. When the pH of a sample of nonfat milk is acidified, it reaches its isoelectric point, and the protein separates out of the solution. The change in pH disrupts the bonds that hold the tertiary structure together. Adjusting the pH of the mixture causes the casein to solidify so it can be removed. A similar process is used in the making of yogurt, cheeses, and cottage cheese. An enzyme provides the acid for the denaturation of the protein for those products. In this experiment, the mass of a quantity of milk and the mass of isolated casein will be determined. From this data, the percent casein in milk will be calculated.

E. Color Tests for Proteins

Certain tests give color products with amino acids, peptides, and/or proteins. The results of the test can be used to detect certain groups or type of bonds within proteins or amino acids.

Biuret test The biuret test is positive for a peptide or protein with two or more peptide bonds. In the *biuret test*, the blue color of a basic solution of Cu^{2+} turns to a violet color when a tripeptide or larger peptide is present. Individual amino acids and dipeptides do not react with the reagent, and the solution will remain blue (negative).

Ninhydrin test The ninhydrin test is used to detect amino acids and most proteins. In the test, most amino acids produce a blue-violet color. Proline and hydroxyproline give a yellow color.

Xanthoproteic test This test is specific for amino acids that contain an aromatic ring. Concentrated nitric acid reacts with the side chains of tyrosine and tryptophan to give nitro-substituted benzene rings that appear as yellow-colored products.

Lab Information

Time: 2 –3 hr

Comments: In protein color tests, be sure to note the color of the reagent before you add it to the samples.
Concentrated HNO_3 and 10% NaOH are extremely corrosive and damage skin and eyes.
Tear out the report sheets and place them beside the matching procedures.

Related topics: Amino acids, peptide bonds, structural levels of proteins, denaturation of proteins

Experimental Procedures

GOGGLES REQUIRED!

A. Peptide Bonds

Materials: Organic model set

A.1 Make models of glycine and serine. Remove the components of water (H—OH) from an amino and carboxylic acid group to form the dipeptides glycylserine and serylglycine. Draw each of their structures.

A.2 Demonstrate hydrolysis of the dipeptide serylglycine by breaking the peptide bond and adding the components of H_2O. Write a balanced equation for the hydrolysis.

B. Structure of Proteins

Materials: Organic model set

B.1 Make a model of an amino acid. Draw its structure and name it. Share your model with two other students, or groups of students who will make models of other amino acids. Line the models up and write their amino acid structures.

B.2 Form peptide bonds between the amino acids by removing the components of water. Write an equation for the formation of a tripeptide. Use the symbols for the amino acids to write the tripeptide order. Rearrange the amino acids and prepare a different tripeptide. Write the structure and name of the new tripeptide.

C. Denaturation of Proteins

Materials: Test tubes, test tube holder, 10-mL graduated cylinder, 1% egg albumin (or an egg, cheesecloth, and beaker), 10% HNO_3, 10% NaOH, 95% ethyl alcohol, 1% $AgNO_3$ (dropper bottle)

A fresh egg albumin solution can be prepared by mixing the white from one egg with 200 mL of water and filtering the mixture through cheesecloth into a beaker.

Place 2–3 mL of egg albumin solution in each of five test tubes. Use one sample for each of the following tests. Record your observations and give a brief explanation for the results.

C.1 **Heat** Using a test tube holder, heat the egg albumin solution over a low flame. Describe any changes in the solution.

C.2 **Acid** Add 2 mL of 10% HNO_3.

C.3 **Base** Add 2 mL of 10% NaOH.

C.4 **Alcohol** Add 4 mL of 95% ethyl alcohol. Mix.

C.5 **Heavy metal ions** Add 10 drops of 1% $AgNO_3$.

D. Isolation of Casein (Milk Protein)

Materials: 150-mL beaker, hot plate or Bunsen burner, thermometer, funnel or Büchner filtration apparatus, filter paper, watch glass, nonfat milk, 10% acetic acid, dropper, pH paper, stirring rod

D.1 Weigh a 150-mL beaker. Add about 20 mL of nonfat milk to the beaker and weigh. Calculate the mass of the nonfat milk sample.

D.2 Using pH indicator paper, determine the pH of the milk sample.

D.3 Warm the sample on a hot plate or a Bunsen burner until the temperature reaches about 50°C. Remove the beaker and milk from the heat and add 10% acetic acid, drop by drop. You may need 2–3 mL. Stir continuously. At the isoelectric point, the casein (milk protein) becomes insoluble. When no further precipitation occurs, stop adding acid. If the liquid layer is not clear, heat the mixture gently for a few more minutes. Determine the pH at which the casein becomes insoluble in solution. This is the pH of the isoelectric point of casein.

D.4 Collect the solid protein using a funnel and filter paper or the Büchner filtration apparatus. Wash the protein with two 10-mL portions of water. Weigh a watch glass. Transfer the protein to the watch glass and let the protein dry. Weigh. Calculate the mass of milk protein. *Save for part E.*

D.5 Calculate the percentage of casein in the nonfat milk.

$$\% \text{ Casein } = \frac{\text{mass (g) of casein}}{\text{mass (g) of milk}} \times 100\%$$

E. Color Tests for Proteins

Materials: Test tubes, test tube rack, 10-mL graduated cylinder, boiling water bath, cold water bath, pH paper, spatula, dropper bottles of 1% amino acid solution (glycine, tyrosine), dropper bottles of 1% solutions of proteins (gelatin, egg albumin), casein from part D, 0.2% ninhydrin solution, concentrated HNO_3 (dropper bottle), 10% NaOH, 5% $CuSO_4$ (biuret), red litmus paper

E.1 **Biuret test** In four separate test tubes, place 2 mL of solutions of glycine, tyrosine, gelatin, and egg albumin. To the fifth tube, add a small amount of the solid casein (from part D), the amount held on the tip of a spatula. To each sample, add 2 mL of 10% NaOH and stir. Then add 5 drops of biuret reagent (5% $CuSO_4$), and stir. Record the color of each sample. The formation of a pink-violet color indicates the presence of a protein with two or more peptide bonds. If such a protein is not present, the blue color of the cupric sulfate will remain (negative result). Record the results and your conclusions.

E.2 **Ninhydrin test** In four separate test tubes, place 2 mL of the solutions of glycine, tyrosine, gelatin, and egg albumin. To the fifth tube, add a small amount of the solid casein (from part D), the amount held on the tip of a spatula. Add 1 mL of 0.2% ninhydrin solution to each sample. Place the test tubes in a boiling water bath for 4–5 minutes. Look for the formation of a blue-violet color. Record your observations.

E.3 **Xanthoproteic test** *(optional)* Place 1 mL of the solutions of glycine, tyrosine, gelatin, and egg albumin in four test tubes. To a fifth tube, add a small amount of casein (from part D). *Cautiously* add 10 drops of concentrated HNO_3 to each sample. Place the test tubes in a boiling water bath and heat for 3–4 minutes. Remove the test tubes, place them in cold water, and let them cool. Carefully add 10% NaOH, drop by drop, until the solution is just basic (turns red litmus blue). This may required 2–3 mL of NaOH. *Caution: heat will be evolved.* Look for the formation of a yellow-orange color, which may vary in intensity. Record your observations.

Report Sheet - Lab 37

Date _____   Name _____

Section _____   Team _____

Instructor _____

Pre-Lab Study Questions

1. What is a peptide bond?

2. How does the primary structure of proteins differ from the secondary structure?

A. Peptide Bonds

A.1   Structure of glycylserine

Structure of serylglycine

A.2   Hydrolysis of serylglycine

Report Sheet - Lab 37

B. Structure of Proteins

B.1   Amino acid structures and names

B.2   Equation for the formation of the first tripeptide

Order of amino acids using symbols

Equation for the formation of the second tripeptide

Order of amino acids using symbols

Report Sheet - Lab 37

C. Denaturation of Proteins

Treatment	Observations of Egg Albumin	Explanation
C.1 Heat		
C.2 Acid		
C.3 Base		
C.4 Alcohol		
C.5 Heavy metal ions		

Questions and Problems

Q.1 Why are heat and alcohol used to disinfect medical equipment?

Q.2 Why is milk given to someone who accidentally ingests a heavy metal ion such as silver or mercury?

D. Isolation of Casein (Milk Protein)

D.1 Mass of beaker	
Mass of beaker and milk	
Mass of milk	
D.2 pH of milk	
D.3 pH when casein precipitates	
D.4 Mass of watch glass	
Mass of watch glass and casein	
Mass of casein	
D.5 Percent casein *Show calculations.*	

Questions and Problems

Q.3 Compare the pH of the milk sample and the pH at which the casein solid forms.

Q.4 How does a change in pH affect the structural levels of a protein?

Report Sheet - Lab 37

E.   Color Tests for Proteins

Observations of Color Tests			
Sample	E.1 **Biuret**	E.2 **Ninhydrin**	E.3 **Xanthoproteic**
Glycine			
Tyrosine			
Gelatin			
Egg albumin			
Casein (milk protein)			

Questions and Problems

Q.5   After working with HNO_3, a student noticed that she had a yellow spot on her hand. What might be the reason?

Q.6   Which samples give a negative biuret test? Why?

Q.7   What functional group gives a positive test in the xanthoproteic test?

Q.8   What tests could you use to determine whether an unlabeled test tube contained an amino acid or a protein?

Goals

- Prepare a solution of the enzyme amylase.

- Describe the role of an enzyme as a catalyst in biological systems.

- Set up chemical tests that indicate the rate of an enzyme-catalyzed reaction.

- Observe the effects of enzyme concentration, temperature, pH, and inhibitors upon enzyme activity.

Discussion

In biological systems, reactions are catalyzed by enzymes, which speed up reactions while operating at mild temperature and pH. Like all catalysts, enzymes lower the energy of activation needed for a reaction to take place. For example, an enzyme in blood called carbonic anhydrase converts carbon dioxide and water to carbonic acid. The enzyme catalyzes the reaction of about 35 million molecules of CO_2 every minute. See Figure 38.1

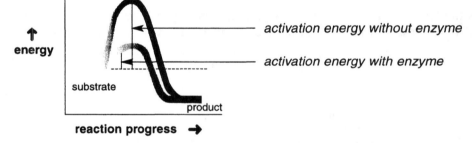

Figure 38.1 Activation energy of a reaction with and without an enzyme (catalyst)

In this experiment you will use amylase, an enzyme that begins the hydrolysis of carbohydrates (amylose) in the mouth. In the presence of amylase, a sample of starch will undergo hydrolysis to give smaller polysaccharides, dextrins, maltose, and glucose.

$$\text{Starch (amylose)} \xrightarrow{\text{Amylase}} \text{Smaller polysaccharides, dextrins, maltose} \xrightarrow{\text{Amylase}} \text{Glucose}$$

Testing Enzyme Activity

We will determine the reaction of amylase with starch by testing for starch using iodine reagent. When iodine is added to a starch solution, a blue-black color is produced. However, if amylase is present and the starch is hydrolyzed, the iodine test is no longer positive. Only the red or gold color of the iodine solution is seen. The faster the amylase breaks down starch, the more quickly the blue-black color is lost. If the blue-black color persists, the enzyme is inactive.

The hydrolyzed product glucose can be detected by using Benedict's reagent. With glucose, the Benedict's reagent turns from blue to a green or reddish-orange with higher glucose concentrations. Table 38.1 summarizes the tests for amylose and glucose.

Table 38.1 *Detection Tests for Starch and Glucose*

Starch (Amylose)	Glucose
Positive iodine test (Turns deep blue with iodine)	Negative iodine test (No color change with iodine. Remains yellow-orange)
Negative Benedict's test (Remains blue)	Positive Benedict's test (Turns green to reddish-orange)

A. Effect of Enzyme Concentration

During catalysis, an enzyme combines with the reactant or *substrate* of a reaction to give an *enzyme-substrate* complex. To form this complex, the substrate fits into the *active site,* where reaction takes place. The *products* are released and the enzyme is ready to catalyze another reaction.

$$E \; + \; S \; \rightleftarrows \; ES \; \rightleftarrows \; E \; + \; P$$

Enzyme	*Substrate*		*Enzyme–substrate complex*		*Enzyme*	*Product*

If the enzyme concentration is increased while substrate concentration is constant, the rate of the reaction will increase. With more enzyme, more substrate molecules can react.

B. Effect of Temperature

The *optimum* temperature is the temperature at which an enzyme operates at maximum efficiency. At low temperatures, the rate of reaction is slowed. At high temperatures, the enzyme protein is denatured. See Figure 38.2.

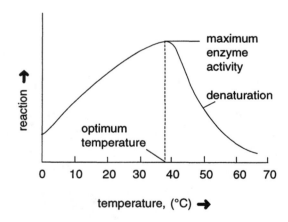

Figure 38.2 Effect of temperature on enzyme activity

C. Effect of pH

At the optimum pH, an enzyme is most active. At pH values above and below optimum, the protein structure of the enzyme is altered, which can severely reduce the enzyme's activity. See Figure 38.3.

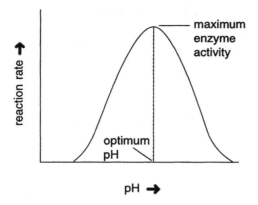

Figure 38.3 Effect of pH on enzyme activity

D. Inhibition of Enzyme Activity

Substances that limit or stop the catalyzing activity of an enzyme are called *inhibitors*. A *competitive inhibitor* blocks the active site of an enzyme, while a *noncompetitive inhibitor* binds to the surface of the enzyme and disrupts the structure of the active site. An *irreversible inhibitor* forms bonds with side-chains of the amino acids in the active site, which makes the enzyme inactive.

Lab Information

Time: $2^{1}/_{2}$ hr
Comments: Tear out the report sheets and place them beside the matching procedures.

Related topics: Enzymes, active site, lock-and-key theory, enzyme activity, factors affecting enzyme activity, inhibition, denaturation

Experimental Procedures

WEAR YOUR SAFETY GOGGLES!

Constant temperature water baths: If commercial water baths are not available, prepare your own. Fill three 250-mL or 400-mL beakers about half full of tap water. Warm one to about 37°C, and heat one to boiling. To the third, add ice to give an ice bath with a low temperature (0-5°C)

Type of Bath	Temperature of Water Baths
Boiling hot water bath	100°C
Warm water	37°C (body-temperature)
Ice bath	0° (or lower than 5°C)

Starch solution: Place 50 mL of 1% starch solution in a small beaker.

Iodine reagent: Obtain a small amount of iodine reagent in a second small beaker. Dropper bottles of iodine may be available. In a third small beaker, place 40 mL of a freshly prepared amylase solution. Obtain a spot plate or plastic sheet for testing. Use clean droppers and rinse. Rinsing the droppers is important to avoid transferring enzyme to other samples.

Amylase preparation: Your instructor may provide a commercial amylase solution.

Reference Tests

Starch: Place a few drops of starch solution in a depression in the spot plate. Add 1 drop of iodine reagent. The reaction with starch should give a deep blue-black color.

Glucose sample: Place 3 mL of 1% glucose in a test tube. Add 2 mL of Benedict's solution and heat for 3–4 minutes in a boiling water bath. The reaction should give a solid with a red-orange color.

Visual Color Reference

As you proceed with each experiment, you will check enzyme activity by adding iodine to the starch mixture with iodine. When enzyme activity is high, the time required for the starch to hydrolyze will be very short. When the enzyme is slowed down or inactive, the blue-black color will be seen for a longer time. By observing the disappearance of starch, you can assess the relative amount of enzyme activity as follows:

Iodine Test for Starch	Amount of Starch Remaining	Enzyme Activity Level	
Dark blue-black	All	None	0
Blue	Most	Low	1
Light brown	Some	Moderate	2
Gold	None	High	3

A. Effect of Enzyme Concentration

Materials: Test tubes, test tube rack, thermometer, 37°C water bath, droppers, spot plate (or plastic sheet), amylase preparation, 1% starch (buffered pH 7.0), iodine reagent, Benedict's solution, boiling water bath, 5- or 10-mL graduated cylinder

A.1 Fill a large beaker about 2/3 full of tap water. Warm to about 37°C and try to maintain the temperature. Your lab may have a commercial water bath set at 37°C.

Place 4 mL of 1% starch in each of four separate test tubes labeled 1–4. Place 4 mL of amylase solution in a fifth tube. Place all the tubes in a 37°C water bath for 5 minutes. Tube 1 without enzyme is the control. Add 3 drops of the warmed amylase solution to the second tube, 6 drops to the third tube, and 10 drops to the fourth tube. Mix quickly and return the test tubes immediately to the 37°C water bath. Record the time at which you add enzyme. Keep the temperature of the water bath close to 37°C, adding water as needed.

Test Tube	Amylase Solution
1	0 (control)
2	3 drops
3	6 drops
4	10 drops

Immediately, transfer four drops of each reaction mixture (use clean droppers) to a spot plate (or plastic sheet). Add 1 drop of iodine reagent to each. Record your observations for each sample. Use the visual color reference to assess the enzyme activity. Rinse the spot plate or use a new section of plastic sheet. Repeat the test again in 5 and 10 minutes.

A.2 (*Optional*) In the test tubes where hydrolysis occurred, the presence of glucose can be confirmed using Benedict's test. Add 3 mL of Benedict's solution to each test tube and place them in a boiling water bath for 3–4 minutes. The appearance of a green to orange-rust color indicates the presence of glucose. If the solution remains clear blue, no hydrolysis of the starch has taken place.

A.3 Make a graph comparing the enzyme activity and the amount of amylase solution at 10 minutes.

B. Effect of Temperature

Materials: Test tubes, test tube rack, test tube holder, droppers, 5- or 10-mL graduated cylinder, beakers for water baths: ice (0°C), warm (37°C), and boiling (100°C), amylase preparation, spot plate (or plastic sheet), 1% starch, iodine reagent

B.1 Place 4 mL of 1% starch solution in each of three test tubes. Place one tube in a boiling water bath, one tube in the 37°C water bath, and one tube in the ice bath. Add 4 mL of amylase solution to three other test tubes and place one each in the three water baths. Let them remain in the water baths for about 10 minutes to allow the solutions to reach the bath temperature.

For each temperature bath, proceed as follows:
a. Record the temperature of the bath. Remove the test tubes, pour the contents together in one tube, mix, and return the mixture to the same temperature bath.

b. After 10 minutes transfer four drops of the mixture to a spot plate (or plastic sheet). Add 1 drop of iodine to each sample. Record the color and activity level.

B.2 Make a graph of the enzyme activity level and the temperature.

C. Effect of pH

Materials: Test tubes, test tube rack, amylase preparation, 37°C water bath, buffers (pH 2, 4, 7, 10), spot plate (or plastic sheet), 1% starch, iodine reagent

C.1 Place 4 mL of buffer solutions of pH 2, 4, 7, and 10 in separate test tubes. Add 4 mL of amylase solution in each. In four other test tubes place 4 mL of 1% starch solution. Place all of the test tubes in a 37°C water bath for about 5 minutes. Pour each of the 1% starch solutions into a separate buffer–amylase tube. Mix and return the mixtures to the 37°C water bath.

After 15 minutes, remove four drops of each reaction mixture (use clean droppers each time) and place in the spot plate or on the plastic sheet. Add 1 drop of iodine reagent to each. Record your observations and the enzyme activity level for each reaction mixture.

C.2 Make a graph of the enzyme activity level and pH.

D. Inhibition of Enzyme Activity

Materials: Test tubes (6), test tube rack, 5- or 10-mL graduated cylinder, amylase preparation, 37°C water bath, solutions in dropper bottles: 1% NaCl, 1% AgNO$_3$, 1% Pb(NO$_3$)$_2$, 1% HgCl$_2$, 95% ethanol, 1% starch, iodine reagent, spot plate (or plastic sheet)

In one test tube, place 4 mL of amylase solution and 10 drops of 1% NaCl. In a second test tube, place 4 mL of amylase solution and 10 drops of 95% ethanol. In a third test tube, place 4 mL of amylase solution and 10 drops of one of the inhibitors that your instructor assigns: 1% AgNO$_3$, 1% Pb(NO$_3$)$_2$, or 1% HgCl$_2$.

Place 4 mL of 1% starch in three other test tubes. Place all six tubes in a 37°C water bath for 5 minutes. Then combine the starch solutions with the amylase solutions, mix, and return the tubes to the 37°C bath for 15 minutes.

After 15 minutes, remove four drops of each reaction mixture (use clean droppers each time) and place in the spot plate or on the plastic sheet. Add 1 drop of iodine reagent to each. Record your observations and the enzyme activity level for each of the reaction mixtures.

Report Sheet - Lab 38

D. Inhibition of Enzyme Activity

Color Produced by Iodine Test and Activity Level					
NaCl	Level	Ethanol	Level	Inhibitor _____	Level

Questions and Problems

Q.8 In which reaction mixture(s) did hydrolysis of starch occur?

Q.9 What substances added to the mixture were inhibitors?

Q.10 How might those substances inhibit enzyme action?

Q.11 What are some differences and/or similarities in the type of inhibition caused by heat, acid or base, and heavy metal ions on enzyme activity?

Goals

- Identify a vitamin as water or fat soluble.

- Compare the vitamin C content in a variety of citrus juices.

- Determine the effect of heat upon vitamin C.

Discussion

A. Solubility of Vitamins

Vitamins are organic compounds required as cofactors or coenzymes for certain enzymes. However, vitamins are not synthesized by the body and must be provided in our diet. The deficiency of a vitamin in the diet can affect the activity of an enzyme, bringing about a deficiency disease. A lack of vitamin C can cause scurvy, a condition characterized by bleeding gums and anemia, and a deficiency in vitamin A is associated with night blindness. A diet low in vitamin D can cause rickets in children, and a deficiency in B_{12} can lead to anemia. See Table 39.1. Because vitamins B and C have polar groups, they are soluble in water. The body usually excretes any excess in these vitamins. Vitamins A, D, E, and K are nonpolar, fat-soluble vitamins.

Table 39.1 *The Water-Soluble Vitamins as Coenzymes*

Vitamin	Coenzyme Function	Deficiency Symptoms
B_1 Thiamine	Decarboxylation of α-keto acids	Beriberi (fatigue, anorexia, nerve degeneration, heart failure)
B_2 Riboflavin	Biological oxidation	Dermatitis, glossitis (tongue inflammation), cataracts
B_3 Niacin	Biological oxidation	Pellagra (scaly skin, muscle fatigue, diarrhea, mouth sores, mental disorder)
B_6 Pyridoxine	Transamination reactions	Dermatitis, fatigue, anemia, irritability
B_{12} Cobalamin	Transfer of methyl groups in biosynthesis of red blood cells, choline, purines	Pernicious anemia, malformed red blood cells, mental disorders
C Ascorbic acid	Synthesis of collagen, protein metabolism, iron absorption, healing of wounds	Scurvy (bleeding gums, slow-healing wounds, muscle pain, anemia)
Biotin	Carboxylation of pyruvic acid to oxaloacetic acid; fatty acid and amino acid metabolism	Dermatitis, fatigue, anemia, nausea, depression

B. Standardization of Vitamin C

The body uses vitamin C, also called ascorbic acid, to fight infection and repair damaged tissues. Vitamin C is found in many fresh fruits and vegetables including oranges, grapefruit, broccoli, lettuce, and green peppers. Because vitamin C is a reducing agent, we can detect it by its reaction with iodine, I_2. Vitamin C is oxidized and the I_2 is reduced. The indicator used in this reaction is a starch solution. When iodine (I_2) is present, the starch turns a deep-blue color. However, when vitamin C is in the solution, the I_2 is reduced to iodide (I^-), and the starch does not form the blue-black color.

| Vitamin C (Ascorbic acid) | Blue-black with starch | Oxidized vitamin C | No color with starch |

C. Analysis of Vitamin C in Fruit Juices and Fruit Drinks

In this experiment, we will determine the number of milligrams of vitamin C that react with 1 mL of iodine reagent. Then using starch as an indicator, we will determine the milligrams of vitamin C present in a sample of fruit juice or drink.

D. Heat Destruction of Vitamin C

Vitamin C loses its activity when it is heated. One sample of juice will be heated to determine the amount of vitamin C destroyed by heat.

Lab Information

Time: $2^1/_2$–3 hr
Comments: Iodine solution will stain books and clothing. Use carefully.
                     Tear out the report sheets and place them beside the matching procedures.
Related topics: Vitamins, enzymes, cofactors, oxidation, titration

Experimental Procedures

SAFETY GOGGLES MUST BE WORN!

A. Solubility of Vitamins

> **Materials:** Test tubes, test tube rack, dropper, methylene chloride (CH_2Cl_2)
> Vitamins A, B, C, D, E, folic acid, or others

Place a small amount of each vitamin provided in the lab in a separate test tube. Add 2 mL of water to each sample. Mix. Record your observations. Are there separate layers? If two layers form, the vitamin is not water soluble. If the vitamin dissolves to give a clear solution, it is water soluble. Record your observations. Test new samples of any of the vitamins not soluble in water by adding 20 drops of methylene chloride (CH_2Cl_2) to those vitamins. Does the vitamin dissolve in the organic solvent? Record your observations.

B. Standardization of Vitamin C

Materials: Vitamin C (100-mg tablet), mortar and pestle, 250-mL Erlenmeyer flask, 50-mL graduated cylinder, 50-mL buret, buret clamp, small funnel, small beaker, 0.1 M HAc, 1% starch, iodine solution, dropper

B.1 Obtain a vitamin C tablet (100-mg vitamin C). If not 100 mg, record the amount (mg) of vitamin C in the tablet as stated on the label. Crush the tablet and transfer it to a 250-mL Erlenmeyer flask. Add 50 mL of distilled water, 2 mL of 0.1 M HAc (acetic acid), and mix. Add 10 drops of 1% starch solution.

B.2 Place a 50-mL buret in a buret clamp on the ring stand. Carefully fill the buret just above the zero mark with the iodine solution. Drain the iodine down to the zero mark. Record the initial level of the iodine solution. See Figure 39.1.

> ## Caution: Keep iodine reagent away from clothes and skin.

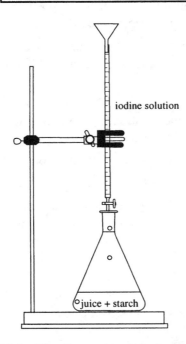
iodine solution

juice + starch

Figure 39.1 Titration setup for vitamin C analysis

Begin adding the iodine solution from the buret to the vitamin C solution until the solution has a deep-blue color. This endpoint is reached after all of the vitamin C has been oxidized and the next drop of iodine solution is not reduced. Record the final reading of the iodine solution in the buret. Calculate the volume of iodine solution used in the titration.

B.3 Calculate the mass (mg) of vitamin C that reacts with 1 mL of iodine solution.

$$\frac{\text{Mass (mg) of vitamin C in tablet}}{\text{Volume (mL) of iodine solution}} = \text{mg vitamin C oxidized by 1 mL iodine solution}$$

C. Analysis of Vitamin C in Fruit Juices and Fruit Drinks

Materials: 125-mL Erlenmeyer flasks, funnel, 50-mL buret, buret clamp, small funnel, small beaker, cheesecloth, 50-mL graduated cylinder, 0.1 M HAc (acetic acid), 1% starch indicator, iodine solution

Fruit juices/drinks: orange, grapefruit, drinks such as HiC or Snapple, or powdered drinks such as Tang or Kool-Aid, or vegetables pureed in a blender or juicer

The experiment works best if the juice is colorless or has a light color.

C.1 Obtain a sample of fruit juice or fruit drink for vitamin C analysis. Record the type of juice or drink. If the juice has a lot of pulp (fiber), pour the juice through cheesecloth that covers a funnel. If suspension particles go through, use two layers of cheesecloth. Use a pipet or small graduated cylinder to transfer a 20-mL sample of the juice into a clean 125-mL Erlenmeyer flask. Add 30 mL of distilled water, 2 mL of 0.1 M HAc (acetic acid), and 10 drops of 1% starch indicator.

Powered drink: If you have a powdered fruit drink such as Tang or Kool-Aid, weigh 1.0 g of the powder and place in a 125-mL Erlenmeyer flask. Add 50 mL of water to the powder, 2 mL of 0.1 M HAc, and 10 drops of 1% starch indicator, and mix. The sample is ready to titrate with iodine.

Vegetables: Weigh 10.0 g of a vegetable. Puree the vegetable in a blender or juicer with 20 mL of water. Filter the puree through cheesecloth. Add water to give a total volume of 50 mL. Add 2 mL 0.1 M HAc and 10 drops of 1% starch indicator.

C.2 Record the initial reading of the level of iodine solution in the buret. Place the flask with the juice and starch mixture under the buret and begin adding iodine solution until the indicator just turns a dark blue. Record the final buret reading for the level of iodine solution.

C.3 Calculate the volume of iodine solution used to reach the endpoint of the titration.

C.4 Calculate the mg of vitamin C in the fruit juice sample. Use the value of mg of vitamin C per 1 mL of iodine solution obtained in step B.3.

$$\underline{} \text{ mL iodine solution} \times \frac{\text{mg vitamin C}}{1 \text{ mL iodine solution}} = \text{ mg vitamin C}$$

Repeat the titration with another sample of juice or other vitamin C sample.

D. Heat Destruction of Vitamin C

Materials: 250-mL Erlenmeyer flasks, funnel, cheesecloth, 50-mL buret, buret clamp, small funnel, small beaker, 50-mL graduated cylinder, 0.1 M HAc (acetic acid), 1% starch indicator, iodine solution, Bunsen burner or hot plate, ice-water bath

Fruit juices/drinks: orange, grapefruit, drinks such as HiC or Snapple, or powdered drinks such as Tang or Kool-Aid, or vegetables pureed in a blender or juicer

D.1 Place 20 mL of a juice in part C that had a high vitamin C content in each of two 250-mL Erlenmeyer flasks. Add 50 mL of water to each. Boil one sample for 10 minutes, and the other for 30 minutes. After 10 minutes, remove the first flask and place it in an ice-water bath. Add 2 mL of 0.1 M HAc and 10 drops of starch indicator. Fill a buret with iodine solution and record the initial level. Titrate with iodine solution to the deep-blue endpoint. Record the final level of iodine solution. Repeat with the other sample.

D.2 Calculate the volume of iodine solution used in each titration.

D.3 Calculate the mg of vitamin C present in each of the heated samples.

D.4 Using the value in C.4 of the mg of vitamin C in the juice sample, calculate the mg of vitamin C that were destroyed by heating the juice.

Report Sheet - Lab 39

Date _____   Name _____

Section _____   Team _____

Instructor _____

Pre-Lab Study Questions

1. What is the difference between a water-soluble vitamin and a fat-soluble one?

2. What is the metabolic role of vitamin C?

3. What foods contain large quantities of vitamin C?

4. What disease is associated with a diet lacking in vitamin C?

Report Sheet - Lab 39

A. Solubility of Vitamins

Vitamin	Soluble in Water	Soluble in CH_2Cl_2	Water or Fat Soluble?	Metabolic Function

Questions and Problems

Q.1 Which vitamins are water soluble?

Q.2 Which vitamins are fat soluble?

Q.3 Which vitamins are required?

Report Sheet - Lab 39

B. Standardization of Vitamin C

B.1   Mass of vitamin C (from label) _____mg

B.2   Initial buret reading _____

     Final buret reading _____

     Volume of iodine solution used _____

B.3   mg Vitamin C per 1 mL iodine solution _____ mg /1 mL iodine solution
      (*Show calculations.*)

C. Analysis of Vitamin C in Fruit Juices and Fruit Drinks

C.1   Type of juice or drink _____ _____

C.2   Initial buret reading _____ _____

     Final buret reading _____ _____

C.3   Volume of iodine solution used _____ _____

C.4   mg Vitamin C in the juice sample _____ _____
      (*Show calculations.*)

Questions and Problems

Q.4   Which of the juices that you analyzed had the most vitamin C?

Q.5   If the daily requirement for vitamin C is 75 mg, how many milliliters (or grams) of each sample
      would you need to consume to meet the minimum daily requirement?

Report Sheet - Lab 39

D. Heat Destruction of Vitamin C

Sample used _____

	Boiled 10 Minutes	Boiled 30 Minutes
D.1 Initial buret reading	_____	_____
Final buret reading	_____	_____
D.2 Volume of iodine solution used	_____	_____
D.3 mg Vitamin C in heated sample (*Show calculations.*)	_____mg	_____mg
D.4 mg Vitamin C destroyed	_____mg	_____mg

Questions and Problems

Q.6 Does heating affect the vitamin C content of a fruit juice?

Q.7 If vitamin C tablets are stored in a warm, humid bathroom cabinet, what might happen to the vitamin C content after a while?

Q.8 If you wish to keep most of the vitamin C content of your vegetables, how should you prepare them for dinner?

DNA Components and Extraction

Goals

- Identify the components in DNA: purines, adenine and guanine, phosphate, and deoxyribose.

- Identify the monomer units in DNA.

- Extract and compare DNA samples from a variety of plant and animal cells.

Discussion

In the cells of all living things are molecules called nucleic acids that provide the information for cellular replication and growth by directing protein synthesis. There are two main types of nucleic acids: DNA (deoxyribonucleic acid), which carries the genetic code to each new generation of cells, and RNA (ribonucleic acid), which carries the instructions for a protein to the ribosomes where proteins are synthesized.

Every living organism contains DNA that provides the directions to make the proteins for the characteristics of that organism. The DNA in an onion has a set of directions to make an onion, whereas the DNA in a banana directs the growth of a banana. Each daughter cell is just like the parent cell because the DNA is duplicated in every cell division.

A. Components of DNA

Nucleic acids are polymers of repeating units known as nucleotides. There are millions of nucleotides in a single DNA molecule. RNA molecules are smaller and contain several thousand nucleotides. A nucleotide of DNA consists of a 5-carbon deoxyribose sugar, one of four nitrogenous bases, and a phosphate group. (See Figure 40.1.)

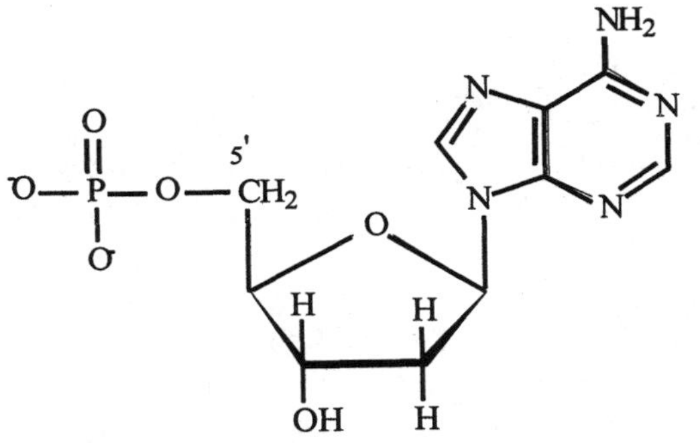

Deoxyadenosine 5'–monophosphate (dAMP)

Figure 40.1 One of four nucleotides found in DNA

In DNA, two bases are the purines adenine (A) and guanine (G) and the other two bases are the pyrimidines cytosine (C) and thymine (T). The nucleotides in RNA differ slightly; the sugar is ribose, and the pyrimidine uracil replaces thymine.

Pyrimidines   Purines

Cytosine (C)   Thymine (T)   Uracil (U)   Adenine (A)   Guanine (G)
(DNA and RNA)   (DNA) only   (RNA only)   (DNA and RNA) (DNA and RNA)

B. Extraction of DNA

In plant cells, DNA strands are combined with protein and RNA molecules. The extraction and isolation of DNA from a cell requires three processes (1) breaking down the cellular membranes, (2) heating the mixture and denaturing the proteins, and (3) precipitating the DNA as a white, stringy, fibrous material.

Step 1 Breaking down cell membranes

The plant or animal material containing the DNA is mixed with an extraction buffer and homogenized to break down the cell membranes. The extraction buffer consists of sodium dodecyl sulfate (SDS), ethylenediamine tetraacetic acid (EDTA), sodium chloride (NaCl), and sodium citrate. SDS disrupts the polar attractions that hold the cell membrane together. EDTA, a chelating agent, removes Mg^{2+} and Ca^{2+} ions that are needed by nucleases that degrade DNA. The NaCl binds with the negatively charged phosphate groups in the DNA fragments, which will cause them to precipitate out of an ethanol solution in step 3. The buffer and the following heat treatment cause lipids and proteins to precipitate out of solution.

Step 2 Heating and denaturing proteins

Heating denatures proteins and inactivates the nuclease enzymes that degrade DNA. If the nucleases remain active, they would hydrolyze DNA into its nucleotide components. When we cook, we heat foods to denature the protein. In addition, some people add a meat tenderizer, which contains a protease, to break down proteins further and make them more digestible.

Step 3: Precipitating DNA from alcohol solution

The DNA is obtained by adding a cold ethanol solution that causes the DNA fragments to come together and precipitate. Because the DNA threads are sticky, they bind to a glass rod so they can be removed from the solution. Following the extraction of DNA, students will make observations and discuss the similarities and differences in DNA samples obtained from different sources.

Lab Information

Time: 2-3 hr
Comments: Tear out the report sheets and place them beside the matching procedures.
Related topics: Deoxyribonucleic acid, nitrogenous bases, deoxyribose, nucleosides, nucleotides, replication

Experimental Procedures

WEAR YOUR PROTECTIVE GOGGLES!

A. Components of DNA

Materials: organic model kits

A.1 Use the C, H, N, and O atoms in a model kit to make a model of one of the purines in DNA nucleotides. Have another lab team make a model of the other purine. Draw the structures of the purines found in the nucleotides of DNA. Save these models.

A.2 Use the C, H, and O atoms in a model kit to make a model of deoxyribose. Draw the structure of deoxyribose. Save this model.

A.3 Use the P, H and O atoms in a model kit to make a model of phosphate. Attach the phosphate group and the purine you constructed to the deoxyribose sugar. Draw the structure of this nucleotide. Write its name. Save this model.

A.4 With your neighbor lab team, combine two nucleotides to form a dinucleotide. Draw the structure of this dinucleotide. Write its name. Combined the two nucleotides to make a different nucleotide. Draw the structure of this dinucleotide.

B. Extraction of DNA

Different student teams may extract DNA from different DNA sources and compare results or the procedure may be run initially with onions and then repeated using a DNA source selected by each student team.

Materials: DNA sources: white onions, variety of other DNA plant sources such as cauliflower, broccoli, garlic, split peas, bananas, and/or animal sources such as chicken liver, calf thymus.
Lab items: knife, two 250-mL beakers, thermometer, hot water bath (400-mL beaker about $^1/_2$ full of water, iron ring, wire screen, and Bunsen burner), blender, cheesecloth or filter paper, microscope
Lab chemicals SDS-NaCl-EDTA buffer solution (Your instructor will prepare this buffer: 50 g sodium dodecyl sulfate, 50 g NaCl, 5 g sodium citrate, 0.2 mL of 0.5 MEDTA and water to make 1 L), isopropanol (placed in an ice bath to keep cold), citrate buffer (0.15 M NaCl, 0.015 M sodium citrate)

1. Obtain about 50 g of a white onion or other DNA source. Use a knife to dice the onion or other DNA source into small pieces. Place the pieces in a 250-mL beaker and add 50 mL of the extraction (SDS-NaCl-EDTA) buffer.

2. Prepare a hot water bath. Place the beaker in the hot water bath, and heat to 60°C. Maintain a temperature of about 60°C by adjusting or removing the flame of the heat source. Allow the beaker to remain in water at 60°C for 15 minutes. (Any longer time in hot water will start to break down DNA.) Stir the mixture occasionally. Remove the beaker and place it in an ice bath for 10 minutes.

Pour the cooled mixture into a blender and blend the contents for 60 seconds using 15-second bursts. Pour the mixture through two pieces of cheesecloth placed over a clean 250-mL beaker. This filtering may take as long as one hour. You may need to use new sets of cheesecloth as it becomes clogged with cell debris.

3. Measure the volume of the filtered onion liquid collected in the beaker and obtain an equal volume of ice-cold isopropanol. If the alcohol is not ice cold, cool it first in an ice bath. Holding the beaker at an angle, slowly pour the alcohol down the side. Allow the solution to sit for 2 minutes. An alcohol layer should form on the top of the filtrate. Because DNA is not soluble in alcohol, the whitish, viscous strands of DNA should precipitate out of the alcohol layer and form a whitish interface. (Other components of the mixture will remain soluble in the alcohol layer.)

Place a glass rod into the solution and turn it slowly to wind the DNA threads into a ball on the end of the glass rod. When you remove the glass rod the DNA will look like a viscous blob. Transfer the DNA from the glass rod to a paper towel and let it dry.

4. Observe the DNA obtained by other student teams with different DNA sources, or repeat the procedure above with another DNA source.

B.1 Observe texture and physical properties of the DNA you have extracted.

B.2 Examine the DNA under a microscope. Record your observations.

Digestion of Foods

Goals

- Identify the types of hydrolysis reactions that take place during the digestion of food.

- Use chemical tests to identify the hydrolysis products of carbohydrates, fats, and proteins.

Discussion

The digestive processes utilize enzymes to carry out the hydrolysis of large food molecules to molecules small enough to dialyze through the intestinal wall into the blood or lymph. See Figure 41.1.

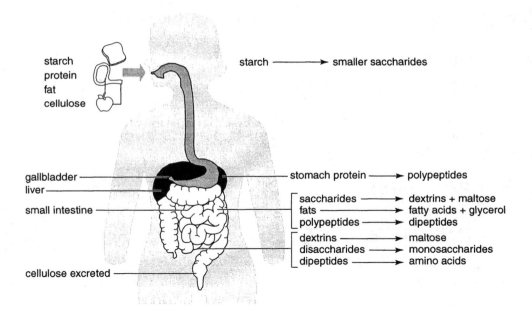

Figure 41.1 Sites of digestion in the human body

A. Digestion of Carbohydrates

Starch, a major carbohydrate in our foods, provides about 50% of our caloric intake. In order to use starch, it must be hydrolyzed into glucose molecules. Digestion of starch begins in the mouth by the action of an enzyme, salivary amylase. Hydrolysis continues in the small intestine through the action of pancreatic amylase, and maltase.

$$\text{Starch (amylose)} \xrightarrow{\textit{Amylase}} \text{maltose} \xrightarrow{\textit{Maltase}} \text{glucose}$$

B. Digestion of Fats

Approximately 25–30% of our diet consists of lipids, primarily fats (triglycerides). Chemically, fat is an ester of glycerol and fatty acids. Digestion of fats begins in the intestine with bile salts and the enzymatic action of lipases obtained from the gallbladder. The bile salts cause the fat to break up into smaller droplets (emulsification), increasing the surface area, and the lipases hydrolyze the ester bonds of the fats.

$$\text{Fats} \xrightarrow{\quad \textit{Pancreatic lipase} \quad} \text{glycerol } + \text{ fatty acids}$$

C. Protein Digestion

Proteins, which make up about 20–25% of our diet, begin to be digested in the stomach where HCl activates the proteases such as pepsin that begin the hydrolysis of peptide bonds. Other enzymes continue to hydrolyze polypeptides and dipeptides.

$$\text{Proteins} \xrightarrow{\quad \textit{Pepsin, chymotrypsin} \quad} \text{peptides, dipeptides} \xrightarrow{\quad \textit{Dipeptidases} \quad} \text{amino acids}$$

Lab Information

Time: 2 hr
Comments: Be careful when you work with boiling water.
                Tear out the report sheets and place them beside the matching procedures.
Related topics: Carbohydrate, fat, protein, hydrolysis, digestion, digestive enzymes

Experimental Procedures

ARE YOUR SAFETY GOGGLES ON?

A. Digestion of Carbohydrates

Materials: Test tubes (2), test tube rack, 10-mL graduated cylinder, droppers, spot plates (or plastic sheets), boiling water bath (large beaker), hot plate or Bunsen burner, amylase solution, 1% starch solution, iodine solution, Benedict's solution

A.1 Hydrolysis of Starch

Amylase preparation: Obtain a spot plate or plastic sheet to test each mixture for starch. In two test tubes, place 4 mL of 1% starch solution. To one, add 10 drops of amylase solution; to the other, add 10 drops of water. Mix thoroughly.

After 5 minutes, use clean droppers to transfer four drops of each mixture to a spot plate (or plastic sheet). Add 1 drop of iodine solution to each. A blue-black color indicates that starch remains in the mixture. Repeat the tests at 10 minutes. Record your observations. Save the mixtures for part A.2.

A.2 Test for Glucose

The presence of glucose as a final product of starch digestion can be determined by adding 3 mL of Benedict's reagent to each of the two test tubes from part A.1. Place the test tubes in a boiling water bath for 5 minutes. Record the colors that form. Benedict's test is positive for glucose if the orange color of Cu_2O forms. Record your observations.

$$\text{Glucose} \quad + \quad 2Cu^{2+} \quad \longrightarrow \quad \text{Gluconic acid} \quad + \quad Cu_2O(s)$$
$$\phantom{\text{Glucose} \quad + \quad} \text{Blue} \phantom{\quad \longrightarrow \quad \text{Gluconic acid} \quad + \quad} \text{Reddish-orange}$$

B. Digestion of Fats

Materials: Test tubes, test tube rack, 50-mL buret, buret clamp, small funnel, small beaker, 25- or 50-mL graduated cylinder, two 250-mL Erlenmeyer flasks, droppers, 37°C water bath, safflower oil, bile salt solution, whole milk, 2% pancreatin, 0.1 M NaOH, pH meter or pH paper, phenolphthalein

B.1 **Action of bile salts** Place 20 drops of safflower oil in each of two test tubes. Add 4 mL water to one test tube. To the other sample, add 2 mL water and 2 mL of bile salt solution. Mix thoroughly. Record your observations of the mixtures in the test tubes. Stir occasionally. After 15 minutes, observe the mixtures in the test tubes again. Look for the separation of layers, or emulsification by the bile salts. Record your observations.

B.2 **Hydrolysis by lipase** Set up a buret containing 0.1 M NaOH. Place 15 mL of whole milk in each of two 250-mL Erlenmeyer flasks. Add 10 mL of 2% pancreatin to each flask and mix thoroughly. Place one flask in a 37°C water bath.

Determine the pH of the milk sample and then titrate the milk sample. Add 3–5 drops of phenolphthalein. Add the 0.1 M NaOH from the buret until a permanent light pink color is obtained. This marks the endpoint. Record the number of milliliters of NaOH used to reach the endpoint.

After 60 minutes, remove the other milk sample from the 37°C water bath. Determine the pH of the sample. Add 3–4 drops of phenolphthalein and titrate with 0.1 M NaOH. Record the number of milliliters of NaOH used to reach the light, permanent pink endpoint.

C. Protein Digestion

Materials: Test tubes (3), test tube rack, 10-mL graduated cylinder, dropper, hard-boiled egg white, 2% pepsin, 0.1 M HCl, 37°C water bath

Cut a piece of egg white into three small pieces about 1 cm in length. Add one piece of egg white to each of three test tubes with the following solutions:

Test Tube	Solutions		
1	4 mL water	+	1 mL (20 drops) 0.1 M HCl
2	4 mL of 2% pepsin	+	1 mL (20 drops) 0.1 M HCl
3	4 mL of 2% pepsin	+	1 mL (20 drops) water

Place the test tubes in a 37°C water bath. Record the appearance of the egg white initially. After 30 minutes, record changes in the egg white in each test tube. Return the test tubes to the water bath and observe again after an additional 30 minutes.

Report Sheet - Lab 41

Date _____ Name _____

Section _____ Team _____

Instructor _____

Pre-Lab Study Questions

1. What type of compound and bond is hydrolyzed by the following?

    a. α-amylase

    b. lipase

2. Why are the reactions of digestion called hydrolysis reactions?

A. Digestion of Carbohydrates

A.1 **Hydrolysis of starch**

Time (minutes)	Starch	Starch + Amylase
5		
10		

A.2 **Test for glucose**

Benedict's solution		
Is glucose present?		

Questions and Problems

Q.1 What is (are) the final product(s) of carbohydrate digestion?

Q.2 Why do some people need lactase when they consume milk products?

B. Digestion of Fats

B.1 Action of bile salts

	Oil Only	Oil + Bile Salts
Initially		
15 minutes		

B.2 Hydrolysis by lipase

Time (minutes)	pH	Volume of 0.1 M NaOH
0		
60		

Questions and Problems

Q.3 What is the function of bile salts in digestion? What organ produces bile salts?

Q.4 Which products of fat hydrolysis would cause a change in pH of the whole milk sample?

Q.5 Write the equation for the action of pancreatin (lipase) on a molecule of tristearin.

Report Sheet - Lab 41

C. Protein Digestion

Appearance of Egg White	Water + HCl	Pepsin + HCl	Pepsin + H_2O
Initial			
After 30 minutes			
After 60 minutes			

Questions and Problems

Q.6 Why does a person with a low production of stomach HCl have difficulty with protein digestion?

Q.7 What are the sites and enzymes that digest protein in the body?

Q.8 What are the final products of protein digestion?

Analysis of Urine

Goals

- Use tests to determine pH, specific gravity, and the presence of electrolytes and organic compounds in urine-like specimens.

- Use chemical tests to analyze for the presence of proteins, glucose, and ketone bodies in urine-like specimens.

Discussion

A.   Color, pH, and Specific Gravity

Examining a urine specimen can give diagnostic information about the processes occurring within the body. The pH, the amounts of electrolytes, uric acid, and glucose can all lead to conclusions about the functioning of the kidneys and liver and the general state of health of the individual. Typically, a person excretes about 1 liter of urine daily. This volume varies with the amount of liquid intake, the temperature, exercise, and the use of substances such as caffeine.

Urine usually has a pH around 6.0, although this varies considerably with diet and activity and can range from 5 to 9 at different times. Urine normally has a light yellow color derived from the breakdown of bilirubin formed during the destruction of red blood cells. The normal range for specific gravity is 1.005 to 1.030. In this experiment, you will test the pH of a urine-like sample, measure its specific gravity, and note its color.

B.   Electrolytes

Urine is about 96% water. The other 4% consists of waste products being eliminated from the body to maintain proper osmotic pressure, electrolyte levels, and pH. Urine normally contains the inorganic ions Cl^-, HCO_3^-, SO_4^{2-}, PO_4^{3-}, K^+, Na^+, NH_4^+, and Ca^{2+}. The presence of sodium ion can be determined by a flame test. The chloride ion will be detected using silver nitrate. Barium ion will be used to detect the presence of sulfate ion. Phosphate ion will be identified with ammonium molybdate.

C.   Glucose

Glucose is not normally detected in the urine although the kidneys do excrete very small amounts of glucose. However, when the glucose level in the blood exceeds the renal threshold, glucose may show up in urine (glucosuria). Concentrations as low as 0.1 g/dL could be abnormal and indicate diabetes mellitus or liver damage. In severe diabetes, glucose levels may reach 5–10 g/dL urine. Reagent strips such as Clinitest or Multistix are used at home and in hospitals to determine the glucose levels in urine. On the strip, glucose oxidase converts glucose to gluconic acid and hydrogen peroxide. A peroxidase enzyme catalyzes the reaction of the hydrogen peroxide with a potassium iodine complex to give colors that range from green to brown.

D. Ketone Bodies

Ketone bodies such as acetone and acetoacetic acid are normally not present in urine. However, they do appear in urine in cases of low-carbohydrate diets, starvation, diabetes mellitus, and liver damage. Reagent test strips such as Ketostix or Multistix are used to detect ketone bodies. The test is positive when acetoacetic acid or acetone in the urine reacts with nitroprusside on the strip. The reaction produces a purple color that deepens in intensity with larger amounts of the ketone bodies.

E. Protein

Normally no protein is detected in urine, although the kidney excretes very small amounts. Typically a person excretes 50–100 mg protein in one day. However, in *proteinuria,* urinary protein levels higher than normal may indicate disease or damage to the kidneys or urinary tract. Protein can be detected by heating a portion of the urine specimen to coagulate the protein. Albustix or Multistix test strips can be used to detect protein in urine. The development of a yellow-green to blue-green color is a positive reaction for protein.

F. Urobilinogen

Urobilinogen is a bile-related pigment that is excreted at low levels (0.1 to 1 Ehrlich units/dL urine). A reagent strip produces a range of brownish-orange colors when the urobilinogen level is elevated (2 Ehrlich units/dL or greater).

Lab Information

Time:	2 hr
Comments:	Be sure to read the test strips at the indicated time after immersing the strip in the fluid. Tear out the report sheets and place them beside the matching procedures.
Related topics:	Electrolytes, pH, ketone bodies

Experimental Procedures

WEAR YOUR SAFETY GOGGLES!

A. Color, pH, and Specific Gravity

> **Materials:** "Urine" specimens (normal and abnormal, prepared by instructor), pH paper or pH meter, urinometer (hydrometer) *or* a small graduated cylinder and beaker

Obtain 20 mL of a "normal urine" specimen and 20 mL of an "abnormal urine" sample. Your instructor will prepare all of the urine specimens. *Do not collect your own urine.* Describe the color of each specimen. Use pH paper or a pH meter to determine the pH of the urine samples. Determine the specific gravity of the urine sample with a urinometer. A display of urinometers may be set up in the lab. If a urinometer is not available, weigh a small beaker. Add 5 mL of the urine sample, and reweigh. Calculate the density and specific gravity of each sample.

B. Electrolytes

Materials: Test tubes, test tube rack, "urine" samples, flame test wire, Bunsen burner, small beakers, 3 M HCl, 0.1 M $AgNO_3$, 0.1 M $BaCl_2$, 3 M HNO_3, ammonium molybdate, warm water bath (70°C)

Sodium ion, Na^+ Dip a cleaned flame test wire (clean in 3 M HCl) into each "urine specimen" and place the wire loop in a flame. A bright, yellow-orange flame indicates the presence of sodium ion.

Chloride ion, Cl^- Place 3 mL of each "urine specimen" in a separate test tube. Add 5 drops of 3 M HNO_3, and 5 drops of 0.1 M $AgNO_3$. A white precipitate (AgCl) confirms the presence of chloride.

Sulfate ion, SO_4^{2-} Place 3 mL of each "urine specimen" in a separate test tube. Add 5 drops of 3 M HCl, and 5 drops of 0.1 M $BaCl_2$. A white precipitate ($BaSO_4$) confirms the presence of sulfate.

Phosphate ion, PO_4^{3-} Place 3 mL of each urine sample in a separate test tube. Add 5 drops of 3 M HNO_3 and 10 drops of ammonium molybdate solution. Place the test tubes in a warm water bath (70°C). A cloudy yellow precipitate confirms the presence of phosphate.

C. Glucose

Materials: Test tubes (2), test tube rack, "urine" samples, Benedict's solution, boiling water bath Reagent strips: Clinistix, or multitest strips such as Multistix, Labstix, or Uristix

Reagent strips: Obtain Clinistix or a multitest strip and dip into a "urine specimen." Compare the test area with the color chart on the container or box. Report your observations. Give a quantitative evaluation of the glucose concentration.

Lab test (optional): A laboratory test for glucose uses Benedict's reagent. Place 10 drops of each "urine" sample in a separate test tube. Add 5 mL Benedict's reagent to each. Place the test tubes in a boiling water bath for 5 minutes. Cool. Record any changes in color. If glucose is present, estimate the amount.

Color with Benedict's Reagent	g/dL
Blue	<0.10
Blue-green	0.25
Green	0.50
Yellow	1
Orange	>2

D. Ketone Bodies

Materials: "Urine" samples; Ketostix, Labstix, or Multistix strips

Reagent strips: Obtain Ketostix, Labstix, or Multistix strips to test each urine specimen for ketone bodies. Dip the reagent strip into a "urine specimen." At 15 seconds, compare the test area with the color chart on the container or box. Report your results as negative or positive. If positive, indicate a small, moderate, or large amount of ketone bodies.

E. Protein

Materials: "Urine" samples; Albustix, Labstix, or Multistix strips; test tubes (2); test tube rack; 1 M HAc; test tube holder

Reagent strips: Obtain Albustix, Labstix, or Multistix for testing protein in the "urine" samples. Dip the strip into a urine specimen. Compare the test areas with the color chart on the container or box. Report your observations.

Lab test (optional): Place 5 mL of each urine sample in separate test tubes. Heat the solution to boiling for 1–2 minutes. If a precipitate forms, add 5 drops of 1 M HAc. Heat for 1 more minute. The formation of a white cloudy precipitate indicates the presence of protein. Report your observations.

F. Urobilinogen

Materials needed: "Urine" samples; Urobilistix, Labstix, or Multistix strips

Reagent Strips: Obtain two Urobilistix, Labstix, or Multistix for testing urobilinogen. Following directions on the package, dip the strip into the "urine specimens." Compare the test areas with the color chart on the container or box. Report your observations.

Appendix
Materials and Solutions

Standard Laboratory Materials

The equipment and chemicals listed throughout the appendix are the materials needed in the laboratory to perform the experiments in this lab manual. The amounts given are recommended for 20–24 students working in teams. The following equipment is expected to be in the laboratory lockers or available from the laboratory stock and will not be listed in each experiment.

Aspirators
Beakers (50–400 mL)
Balances (top loading or centigram)
Büchner filtration apparatus and filter paper
Bunsen burners
Burets (50 mL)
Buret clamps
Clay triangles
Containers for waste disposal
Crucible and cover
Distilled water (special faucet or containers)
Droppers
Evaporating dish
Flask, Erlenmeyer (125–250 mL)
Filter paper for funnels
Funnel
Glass stirring rods
Gloves
Graduated cylinders (5–250 mL)

Hot plates
Ice
Iron rings
Litmus paper
Meter sticks
pH paper
Ring stand
Rulers
Shell vials
Stirring rod, glass
Spatulas
Stoppers
Test tubes (6", 8")
Test tube rack
Thermometer
Tongs (crucible and beaker)
Watch glass
Wire gauze

Additional Materials Needed for Individual Experiments

The equipment and chemicals listed in this section must be supplied by the instructor or stockroom.. The amounts given are recommended for 20–24 students working in teams of two students

1   Measurement and Significant Figures
   A. Measuring Length
      10   Lengths of string
   B. Measuring Volume
      3   Graduated cylinders partially filled with water and a few drops of food coloring
      10   Metal solids
   C. Measuring Mass
      10   Unknown mass samples

2   Conversion Factors in Calculations
   A. Rounding Off
      10   Solid objects with regular shapes

    D. Conversion Factors for Volume

      3   1-liter graduated cylinders   3   1-quart (or two 1-pint) measures

    E. Conversion Factors for Mass

      5   Commercial products with mass given on label in metric and U.S. system units

    F. Percent by Mass

      100 g   Sucrose (sugar)

    G. Converting Temperature

      200 g   Rock salt

3   **Density and Specific Gravity**

    A. Density of a Solid

      10   Metal objects (cubes or cylinders of aluminum, iron, lead, tin, zinc, etc.)

      10   Large graduated cylinders to fit metal objects

      10   Strings or threads (short lengths to tie around metal solids)

    B. Density of a Liquid

      200 mL Isopropyl alcohol   200 mL Corn syrup

      200 mL 20% NaCl or 20% $CaCl_2$

      200 mL Mineral water, vegetable oil, milk, juices, soft drinks, window cleaners, etc.

    C. Specific Gravity

      3–4   Hydrometers in graduated cylinders containing water and the liquids used for (B).

    D. Graphing Mass and Volume

      50   Pennies (pre-1980 or post-1980) or metal pieces (zinc, lead shot, aluminum pellets)

4   **Atomic Structure**

    A. Physical Properties of Elements

           Display of elements (metals and nonmetals)

    B. Periodic Table

           Periodic tables, colored pencils

           Display of elements

5   **Electronic Configuration and Periodic Properties**

    A. Flame Tests

      10   Spot plates   10   Flame test wires

      10   Corks   200 mL 1 M HCl

      100 mL   Unknown solutions (2–3) in dropper bottles (same as test solutions)

      Place in dropper bottles:

      100 mL   0.1 M $CaCl_2$ 100 mL   0.1 M KCl

      100 mL   0.1 M $BaCl_2$ 100 mL   0.1 M $SrCl_2$

      100 mL   0.1 M $CuCl_2$ 100 mL   0.1 M NaCl

6   **Nuclear Radiation**

    A. Background Count

      1   Geiger-Müeller radiation detection tube

    B. Radiation from Radioactive Sources

      1   Geiger-Müeller radiation detection tube

      3–4   Radioactive sources of alpha- and beta-radiation

           Consumer products: Fiestaware®, minerals, old lantern mantles containing thorium compounds, camera lenses, old watches with radium-painted numbers on dials, fertilizers with P_2O, smoke detectors containing Am-241, anti-static devices for records and film containing Po-210, some foods such as salt substitute (KCl), cream of tartar, instant tea, instant coffee, dried seaweed

C. Effect of Shielding, Time and Distance
 1   Geiger-Müeller radiation detection tube
 5–6 Shielding materials such as lead sheets, paper, glass squares, cloth, cardboard
 3–4 Radioactive sources used in 5B.

7 Compounds and their Formulas
 Merck Index or *CRC Handbook of Chemistry and Physics*
 B. Ionic Compounds and Formulas
 $NaCl(s)$
 C. Ionic Compounds with Transition Metals
 $FeCl_3(s)$
 D. Ionic Compounds with Polyatomic Ions
 $K_2CO_3(s)$
 E. Covalent (Molecular) Compounds
 H_2O
 F. Electron Dot Structures and Molecular Shape
 Molecular model set to build models of molecules or ions.

8 Energy and Specific Heat
 A. Specific Heat of a Metal
 10 Styrofoam cups, covers   10 Metal objects
 B. Measuring the Caloric Value of a Food
  This may be an instructor demonstration.
 10 Aluminum cans   Food samples (chips, cheese puffs, etc.)
 10 Buret clamps
 C. Food Calories
 3–4 Food products with nutrition data on labels

9 Energy and States of Matter
 A. A Heating Curve for Water
 5   Timers
 B. Graphing a Cooling Curve
 10 Freezing-point apparatus
    (Large test tube containing a small amount of Salol (phenylsalicylate)
    and fitted with a two-hole stopper, thermometer, and wire stirrer. *The thermometer*
    is frozen in the Salol until heated.
 C. Energy in Changes of State
 10 Styrofoam cups, covers

10 Chemical Reactions and Equations
 A. Magnesium and Oxygen
 10 Magnesium ribbon (2–3 cm)
 B. Zinc and Copper(II) Sulfate
 20 Pieces of zinc(s) strips, 1 cm square 100 mL 1 M $CuSO_4$
 C. Metals and HCl
 10 Cu pieces and Zn pieces 1 cm square 10 Mg ribbon pieces ~ 2 cm long
 250 mL 1 M HCl

D. Reactions of Ionic Compounds
Place 100 mL each in dropper bottles:

0.1 M CaCl$_2$		0.1	M Na$_3$PO$_4$
0.1 M BaCl$_2$		0.1	M Na$_2$SO$_4$
0.1 M KSCN		0.1	M FeCl$_3$

E. Sodium Carbonate and HCl

20	Wood splints	25 g	Na$_2$CO$_3$(s)
250 mL 1 M HCl			

11 Reaction Rates and Equilibrium
A. Exothermic and Endothermic Reactions

50 g	NH$_4$NO$_3$(s)	50 g	CaCl$_2$(s) anhydrous
10	Styrofoam cups, covers	10	Weighing papers
10	Spatulas		

B. Rates of Reactions

10	Cleaned pieces of magnesium ribbon 2–3 cm (about 0.4 g)
100 ml 1.0 M HCl	100 mL 2.0 M HCl
100 mL 3.0 M HCl	200 mL Vinegar (0.1 M HCl)
50 g	NaHCO$_3$ (or baking soda or Alka-Seltzer)

C. Reversible Reactions
dropper bottle sets

100 mL 0.1 M CuCl$_2$	100 mL 0.1 M NaOH
100 mL 1 M NH$_4$OH	100 mL 0.1 M HCl

D. Iron(III)-thiocyanate Equilibrium

10	100-mL beaker	200 mL	0.01 M Fe(NO$_3$)$_3$,
50 mL	1 M Fe(NO$_3$)$_3$	50 mL	0.01 M KSCN
50 mL	1 M KSCN	50 mL	3 M HCl
50 mL	3 M NaOH		Ice

12 Moles and Chemical Formulas
A. Finding the Simplest Formula

10	Heat resistant pads
10	Mg ribbon (0.2–0.3 g, 16–18 cm strips)
	Steel wool

B. Formula of a Hydrate

10	Heat resistant pads
100 g	Hydrate of MgSO$_4$•7H$_2$O

13 Gas Laws
B. Charles' Law
Stoppers may be prepared by stock room with glass inserted attach to a short piece of rubber tubing

10	One-hole stoppers to fit 125-mL Erlenmeyer flasks		
10	Short pieces of glass	10	Short pieces of rubber tubing
10	Pinch clamps	10	Water pans
20	Boiling chips		

14 Partial Pressures of Oxygen, Nitrogen, and Carbon Dioxide
A. Partial Pressures of Oxygen and Nitrogen in Air

25 g	Fe (iron) filings	1	Barometer

B. Carbon Dioxide in the Atmosphere *(May be done as an instructor demonstration)*
The following items may be assembled and available for use by instructor:

10	Glass tubing (60–75 cm)		
10	Two-hole stoppers with two short pieces of glass tubing		
20	Rubber tubing (1 long, 1 short)	10 mL	Food coloring (optional)
20 mL	Mineral oil	100 mL	6 M NaOH

C. Carbon Dioxide in Expired Air *(May be done as an instructor demonstration)*
The following items may be assembled and available for use:

10	Glass tubing (60–75 cm)		
10	Two-hole stoppers with two short pieces of glass tubing		
20	Rubber tubing (1 long, 1 short)	10	Pinch clamps
20	Clean straws to fit tubing	10 mL	Food coloring (optional)
20 mL	Mineral oil	100 mL	6 M NaOH

15 Solutions, Electrolytes, and Concentration

A. Polarity of Solutes and Solvents *(May be a demonstration)*

20 g	$KMnO_4(s)$	20 g	$I_2(s)$
20 g	Sucrose(s)	200 mL	Cyclohexane
20 mL	Vegetable oil (2 dropper bottles)		

B. Electrolytes and Conductivity

1	Electrical conductivity apparatus (**Instructor use only**)		
100 g	NaCl(s)	100 mL	0.1 M NaOH
100 mL	0.1 M HCl	100 mL	0.1 M $HC_2H_3O_2$
100 mL	0.1 M NH_4OH	100 mL	0.1 M NaCl
100 mL	0.1 M sucrose	100 mL	0.1 M glucose
100 mL	ethanol	100 mL	soft drinks, mineral water, etc.

C. Electrolytes in Body Fluids

3	IV solutions in bottles or bags (1 L, 500 mL, or 100 mL)
	Examples: 5% glucose, 0.95% NaCl, and/or electrolytes eg. saline, Ringer's

D. Concentration of a Sodium Chloride Solution

10	10-mL Graduated cylinders (or pipettes)
200 mL	NaCl solution (saturated)

16 Soluble and Insoluble Salts

A. Soluble and Insoluble Salts

10	Spot plates or 10 transparency sheets,
100 mL	0.1 M solutions (NaCl, Na_2SO_4, $Ba(NO_3)_2$, $AgNO_3$, Na_3PO_4, $CaCl_2$, NaOH, Na_2CO_3)
20	Droppers, or use dropper bottles of the 0.1 M solutions

B. Solubility of KNO_3 *(Work in Teams)*

10	10-mL graduated cylinders	100 g	$KNO_3(s)$
10	Weighing papers	10	Spatulas

C. Testing the Hardness of Water

500 mL	Tincture of green soap
500 mL	Water samples (distilled water, hard water, soft water, tap water, sea, lake or river water, mineral water, rain water, fish tank, pool or hot tub water)

D. Purification of Water

1000 mL	Muddy water	100 mL	1% $Al_2(SO_4)_3$
100 mL	1% Na_2SO_4	100 mL	1% NaCl

17 Testing for Cations and Anions

A. Tests for Positive Ions (Cations)

10	Spot plates	10	Flame test wires
200 mL	3 M HCl	200 mL	6 M HNO_3

100 mL   6 *M* NaOH

Place 100 mL each in dropper bottles:

0.1 *M* NaCl	0.1 *M* KCl	0.1 *M* KSCN
0.1 *M* CaCl$_2$	0.1 *M* (NH$_4$)$_2$C$_2$O$_4$	
0.1 *M* NH$_4$Cl	0.1 *M* FeCl$_3$	

B. Tests for Negative Ions (Anions)

Place 100 mL each in dropper bottles:

0.1 *M* NaCl	0.1 *M* AgNO$_3$
6 *M* HNO$_3$	3 *M* HCl
0.1 *M* Na$_2$SO$_4$	0.1 *M* BaCl$_2$
0.1 *M* Na$_3$PO$_4$	0.1 *M* Na$_2$CO$_3$

(NH$_4$)$_2$MoO$_4$ reagent

50 mL   Unknowns: KCl, Na$_2$CO$_3$, (NH$_4$)$_2$SO$_4$, CaCl$_2$, K$_2$SO$_4$, (NH$_4$)$_3$PO$_4$, etc.

D. Testing Consumer Products for Some Cations and Anions

Samples of consumer products for testing: juice, milk, salt, substitute salt, window cleaner, plant food, baking soda, baking power, antacids, etc.

2–4   Consumer products such as juice, milk, plant food, bone meal, fertilizer, substitute salt (KCl), Epsom salts (MgSO$_4$•7H$_2$O)

18   Solutions, Colloids, and Suspensions

A. Identification Tests

200 mL	1% starch	200 mL	10% glucose
200 mL	10% NaCl	500 mL	Benedict's reagent

Place the following in dropper bottles:

100 mL	0.1 *M* AgNO$_3$	100 mL	Iodine solution

B. Osmosis and Dialysis

10	20-cm dialysis bag (cellophane tubing)		
100 mL	10% NaCl	100 mL	10% glucose
100 mL	1% starch	100 mL	0.1 *M* AgNO$_3$
100 mL	Benedict's solution	100 mL	Iodine solution

C. Filtration

25 g	Powdered charcoal	200 mL	1% starch
100 mL	Iodine solution		

19   Acids, Bases, pH, and Buffers

A. pH Color Using Red Cabbage Indicator

1	Red cabbage	200 mL	Buffers pH 1–13

B. Measuring pH

2–3	pH meters	2–3	Wash bottles
2–3	Boxes Kimwipes	200 mL	Cabbage indicator from part A
20 mL	Buffers (pH 4, pH 10)		

Samples to test for pH: bring from home or have in lab. *Examples:* shampoo, hair conditioner, mouthwash, antacids, detergents, fruit juice, vinegar, cleaners, aspirin

C. Effect of Buffers on pH

2–3	pH meters	2–3	Wash bottles
2–3	Boxes Kimwipes	100 mL	0.1 *M* NaCl
100 mL	0.1 *M* HCl	100 mL	0.1 *M* NaOH
100 mL	pH 4 buffer	100 mL	pH 10 buffer

20   Acid–Base Titration

A. Acetic Acid in Vinegar

10	5-mL pipet and bulbs	200 mL	Vinegar (white)

1L   0.1 *M* NaOH (standardized)   100 mL Phenolphthalein indicator

B. Titration of an Antacid

2–4   Antacid products (Tums, Rolaids, Maalox, etc.)

5   Mortar and pestle   1 L   0.1 *M* HCl (standardized)

100 mL Phenolphthalein

21 Properties of Organic Compounds

A. Color, Odor, and Physical State *(May be a display in lab)*

1–2   Chemistry handbook   20 g   NaCl(s)

20 g   KI(s)   20 g   Benzoic acid(s)

20 mL   Toluene   20 mL   Cyclohexane

B. Solubility *(This may be an instructor demonstration.)*

20 g   NaCl(s)   20 mL   Toluene

20 mL   Cyclohexane

C. Combustion *(This may be an instructor demonstration.)*

30   Wood splints   10 g   NaCl(s)

10 mL   Cyclohexane

D. Functional Groups

10   Organic model kit or prepared models of organic compounds to observe

22 Structures of Alkanes

A. Structures of Alkanes, B. Isomers, C. Cycloalkanes, and D. Haloalkanes

10   Organic model kits or prepared models

2   Chemistry handbooks

23 Reactions of Hydrocarbons

A. Types of Hydrocarbons

10   Organic model kits or prepared models

B. Combustion *(This may be an instructor demonstration.)*

5   Wooden splints   50 mL   Cyclohexane

50 mL   Cyclohexene   50 mL   Toluene

C. Bromine Test *(This may be an Instructor Demonstration)*

50 mL   Cyclohexane   50 mL   Cyclohexene

50 mL   Toluene   50 mL   Unknowns (use same compounds)

100 mL 1% Br_2 in CH_2Cl_2

D. Potassium Permanganate ($KMnO_4$) Test

50 mL   Cyclohexane   50 mL   Cyclohexene

50 mL   Toluene   50 mL   Unknowns (use same compounds)

100 mL 1% $KMnO_4$

E. Identification of Unknown

Unknowns of same substances

24 Alcohols and Phenols

A. Structures of Alcohols and Phenol

10   Organic model kits or prepared models

Use for B. C. and D.

50 mL   Ethanol   50 mL   *t*-butyl alcohol (2-methyl-2-propanol)

50 mL   Cyclohexanol   50 mL   2-propanol

50 mL   20% phenol   2–4   Unknowns (Use same compounds)

C. Oxidation of Alcohols

100 mL 2% chromate solution

D. Ferric Chloride Test

100 mL 1% $FeCl_3$

E. Identification of Unknown
Unknowns of same substances

25 Aldehydes and Ketones
A. Structures of Some Aldehydes and Ketones
10   Organic model kits or prepared models
B. Properties of Aldehydes and Ketones
50 mL	Acetone	50 mL	Benzaldehyde
50 g	Camphor	50 mL	Cinnamaldehyde,
50 mL	Vanillin	50 mL	Propionaldehyde (propanal)
50 mL	Cyclohexanone	50 mL	2,3-Butanedione
2	Chemistry handbooks	2–3	Unknowns (use above compounds)

C. Iodoform Test for Methyl Ketones (Test tubes from part B.3)
Use compounds from B
100 mL 10% NaOH   200 mL Iodine test reagent
D. Oxidation of Aldehydes and Ketones
Use compounds from B   500 mL Benedict's reagent
E. Identification of an Unknown
Unknowns of same substances

26 Types of Carbohydrates
A. Types of Carbohydrates and B. Disaccharides
10   Organic model kits or prepared models

27 Tests for Carbohydrates
A. Benedict's Test for Reducing Sugars
200 mL 2% starch   500 mL Benedict's reagent
Place 50 mL each in dropper bottles:
2% glucose   2% fructose   2% sucrose   2% lactose
50 mL Unknown solutions from above *Examples*: glucose, fructose, sucrose, and lactose
B. Seliwanoff's Test for Ketoses
200 mL 2% starch   100 mL Seliwanoff's reagent
50 mL   Unknown solutions from above *Examples*: glucose, fructose, sucrose, and lactose
Place 50 mL each in dropper bottles:
2% glucose   2% fructose   2% sucrose   2% lactose
50 mL Unknown solutions from above *Examples*: glucose, fructose, sucrose, and lactose
C. Fermentation Test
6   Fermentation tubes (or 6 small test tubes and 6 large test tubes),
200 mL 2% starch   40 g   Baker's yeast (fresh)
50 mL   Unknown solutions from above
Examples: glucose, fructose, sucrose, and lactose
Place 50 mL each in dropper bottles:
2% glucose   2% fructose   2% sucrose   2% lactose
50 mL Unknown solutions from above *Examples*: glucose, fructose, sucrose, and lactose
D. Iodine Test for Polysaccharides
10   Spot plates
200 mL 2% starch   100 mL   Iodine reagent
Place 50 mL each in dropper bottles:
2% glucose   2% fructose   2% sucrose   2% lactose
50 mL   Unknown solutions from above *Examples*: glucose, fructose, sucrose, and lactose
E. Hydrolysis of Disaccharides and Polysaccharides
10   Spot plate (or watch glasses)

100 mL	10% NaOH	100 mL	10% HCl
100 mL	Iodine reagent	500 mL	Benedict's reagent

Place 50 mL each in dropper bottles:

2% glucose   2% fructose   2% sucrose   2% lactose

F. Testing Foods for Carbohydrates

3–4   Sugar samples (refined, brown, "natural," powdered), honey
          Syrups (corn, maple, fruit), Foods with starches: cereals, pasta, bread

100 mL	Seliwanoff's reagent	100 mL	Iodine reagent
500 mL	Benedict's reagent		

28 Carboxylic Acids and Esters

A. Carboxylic Acids and Their Salts

100 mL	10% NaOH	100 mL	10% HCl
50 mL	Glacial acetic acid	50 g	Benzoic acid

B. Esters

10	Organic model sets		
200 mL	Glacial acetic acid	50 g	Salicylic acid
20 mL	Methanol	20 mL	1-Pentanol
20 mL	1-Octanol	50 mL	85% H_3PO_4 (dropper bottle)
20 mL	1-Propanol	20 mL	Benzyl alcohol

C. Hydrolysis of Esters

Place in dropper bottles:

100 mL	Methyl salicylate	100 mL	10% NaOH
100 mL	10% HCl		

29 Aspirin and Other Analgesics

A. Preparation of Aspirin

10	Pans or large beakers	50 g	Salicylic acid(*s*)
100 mL	Acetic anhydride	50 mL	85% H_3PO_4 in a dropper bottle

B. Testing Aspirin Products

3–4   Commercial brands of aspirin and buffered aspirin, purified aspirin (and/or crude
          aspirin)

20 g	Acetylsalicylic acid	100 mL	0.15% Salicylic acid
100 mL	1% $FeCl_3$	100 mL	10% NaOH
100 mL	10% HCl		

C. Analysis of Analgesics

Saran wrap, rubber band to fit beaker		60	Micropipettes
10	Spot plates		

Dropper bottles containing 1% solutions in ethanol of the following:

1% aspirin	1% ibuprofen
1% acetaminophen	1% naproxen
1% caffeine	1% over the counter drugs,

400 mL Solvent (Prepare by combining 300 mL ethyl acetate with 100 mL hexane)

10	TLC plate with silica gel		UV lamp (short wavelength 254 nm)

30 Lipids

A. Triacylglycerols

5   Organic model kits or prepared models

B. Physical Properties of Some Lipids and Fatty Acids

25 g	Lecithin	25 g	Stearic acid
25 g	Cholesterol	100 mL	Methylene chloride, CH_2Cl_2

Place 25 mL each in dropper bottles:

Oleic acid	Vitamin A
Olive oil	Safflower oil

C. Bromine Test for Unsaturation
5   Organic model kits or models   Samples from B
100 mL 1% Br_2 in CH_2Cl_2

D. Preparation of Hand Lotion
Team project: Steps D.1, D.2, and D.3 may be prepared by different teams in the lab.

10	10-mL graduated cylinders	20	50-mL or 100-mL beakers
50g	Stearic acid	15 g	Cetyl alcohol
25 g	Lanolin (anhydrous)	15 mL (dropper)	Triethanolamine
25 mL	Glycerin	100 mL	Ethanol
100 mL	Distilled water		Fragrance (optional)

Commercial hand lotion products

31 Glycerophospholipids and Steroids
A. Isolating Cholesterol from Egg Yolk

10	Eggs	10	50-mL flasks
10	Steam baths	10	100-mL beakers
10	Short-stem funnels, glass wool	2	Melting point apparatus
500 mL	acetone		

B. Isolating Lecithin from Egg Yolk
Egg yolk residue from Part A

10	Steam baths	10	100-mL beaker
2	Melting point apparatus	500 mL	Ethyl ether

32 Saponification and Soaps
A. Saponification: Preparation of Soap
Optional: Hot plate and a stirring bar   200 mL Ethanol
100 mL 20% NaOH   200 mL Saturated NaCl solution
10 pairs of disposable gloves
100 g   Solid fats: lard, coconut oil , solid shortening, coconut oil
100 mL Liquid vegetable oils, olive or other vegetable oil

B. Properties of Soaps and Detergents
50 g   Commercial soaps, lab-prepared soap (from part A), detergent
50 mL   Safflower oil   100 mL 1% $CaCl_2$
100 mL 1% $MgCl_2$ 100 mL 1% $FeCl_3$

33 Amines and Amides
A. Structure and Classification of Amines
10   Organic model kits or prepared models
B. Solubility of Amines in Water
Place in dropper bottles:
30 mL Aniline   30 mL Triethylamine
30 mL *N*-Methylaniline   100 mL 10% HCl
C. Neutralization of Amines with Acids
Test tubes from part B 1   100 mL 10% HCl
D. Amides
5   Organic model kits   50 g   Acetamide
50 g   Benzamide
Place 100 mL each in dropper bottles:

10% NaOH   10% HCl

34 Synthesis of Acetaminophen

A. Synthesis of Acetaminophen

25 g   *p*-Aminophenol   10 mL   Acetic anhydride

100 mL 85% H_3PO_4 in dropper bottle

B. Isolating Acetanilide from an Impure Sample

50 g   Impure acetanilide

35 Plastics and Polymerization

A. Classification of Plastics

Samples of plastic items: Nylon, Teflon tape, Saran, Styrofoam cups, plastic cups, milk cartons, yogurt containers, buttle wrap, detergent bottles, soda bottles, etc.

40   Small pieces of each type of plastic.

100 mL Vegetable oil   100 mL Ethanol

100 mL Glycerin   100 mL Acetone

B. Gluep and Slime®

10   10-mL, 50-mL graduated cylinders

50   Styrofoam cups   20   Plastic sticks or spatulas

          Plastic gloves

500 ml saturated borate solution 800 mL Elmer's glue

200 mL 4 % polyvinyl alcohol solution

C. Polystyrene

10   Funnels   10   Filter papers

10   Hot plates   20   10- or 20-mL beakers

10   50-mL beakers   10   Wood sticks

          Heavy duty aluminum foil

50 g   Alumina   50 mL   Styrene

10 g   Benzoyl peroxide (or an acne preparation which contains 5% or10% benzoyl peroxide),

D. Nylon

10   50-mL, 100-mL beakers   10   Forceps

10   10-mL and 50-mL graduated cylinders   10   Metal spatulas

          Gloves (must not dissolve in hexane)

300 mL 50% aqueous ethanol solution   100 mL 6 M HCl, 6 M NaOH

100 mL Acetone

250 mL Solution 1: 4% hexamethylenediamine and NaOH

          (dissolve 3.0 g $H_2N(CH_2)_6NH_2$ and 1.0 g NaOH in 50 ml of distilled water)

          If solid, place the reagent bottle in hot water to melt (mp 39°C).

250 mL Solution 2: 4% sebacoyl chloride, $ClCO(CH_2)_8COCl$, in hexane

          (dissolve 2.0 ml sebacoyl chloride in 50 ml hexane.)

36 Amino Acids

A. Amino Acids

10   Organic model kits or prepared models

B. Chromatography of Amino Acids

1 box   Plastic wrap   1 box   Whatman #1 filter paper (12 cm × 24 cm)

50   Toothpicks or capillary tubing   1   Drying oven (80°C)

2   Hair dryers (optional)   1   Stapler

Place 50 mL each in dropper bottles:

1% Alanine   1% Glutamic acid

1% Serine   1% Aspartic acid

1% Lysine   1% Phenylalanine

50 mL   1% Unknown amino acids (Use samples from above list)

Chromatography solvent

100 mL 0.5 M NH$_4$OH   200 mL Isopropyl alcohol

0.2% Ninhydrin spray reagent (in ethanol or acetone)

37 Peptide and Proteins

A. Peptide Bonds and B. Structure of Proteins

5   Organic model set

C. Denaturation of Proteins

Place in dropper bottle:

100 mL 1% egg albumin   Dissolve 1 g egg albumin in water to make 100 mL or students can make a fresh egg albumin solution by mixing the egg white from one egg with 200 mL of water and filtering the mixture through cheesecloth into a beaker.

100 mL 10% HNO$_3$ 100 mL 10% NaOH

100 mL 95% ethanol   100 mL 1% AgNO$_3$

D. Isolation of Casein (Milk Protein)

200 mL Nonfat milk   200 mL 10% Acetic acid

E. Color Tests for Proteins

10 g   Casein from part D

Place 100 ml each in dropper bottles:

1% Glycine   1% Tyrosine

1% Gelatin   1% Egg albumin (See part C.)

10%   NaOH   50 mL   Conc. HNO$_3$

1 can   0.2% Ninhydrin reagent   200 mL 5% CuSO$_4$

38 Enzymes

A. Effect of Enzyme Concentration

10   Spot plate (or plastic sheets)

10   Timer   1%   Starch (buffered to pH 7)

200 mL Amylase preparation   100 mL 1% Glucose

100 mL Iodine reagent   500 mL Benedict's reagent

B. Effect of Temperature

Amylase preparation   200 mL 1% Starch

100 mL Iodine test reagent

C. Effect of pH

Amylase preparation   100 mL Buffers (pH 2, 4, 7, 10)

200 mL 1% Starch   100 mL Iodine test reagent

D. Inhibition of Enzyme Activity

Amylase preparation

Place 50 mL each in dropper bottles:

1% NaCl   1% AgNO$_3$

1% CuSO$_4$ 1% Pb(NO$_3$)$_2$

1% HgCl$_2$ Ethanol

200 mL 1% Starch   100 mL Iodine reagent

39 Vitamins

A. Solubility of Vitamins

4–6   Samples of vitamins A, B, C, D, E, folic acid, or others

50 mL   Methylene chloride (CH$_2$Cl$_2$)

B. Standardization of Vitamin C

2   Mortar and pestle   10   Tablets of vitamin C (100-mg)

100 mL 0.1 *M* HAc   100 mL 1% Starch
1 L   Iodine solution

C. Analysis of Vitamin C In Fruit Juices and Fruit Drinks

100 mL 0.1 *M* HC₂H₃O₂   100 mL 1% Starch
1 L   Iodine solution   10   Squares of cheesecloth
         Samples of fruit juices/drinks: orange, grapefruit, drinks such as HiC or Snapple, or powdered drinks such as Tang or Kool-Aid, or vegetables pureed in a blender or juicer

D. Heat Destruction of Vitamin C

10   Squares of cheesecloth   100 mL 1% Starch
100 mL 0.1 *M* HAc   1 L   Iodine solution
3–4   Fruit juices/drinks: orange, grapefruit, drinks such as HiC or Snapple, or powdered drinks such as Tang® or Kool-Aid®, or vegetables pureed in a blender or juicer

40 DNA Components and Extraction

A. Components of DNA

10   Organic model kits

B. Extraction of DNA

Different student teams may extract DNA from different DNA sources and compare results or the procedure may be run initially with onions and then repeated using other DNA sources selected by each student team.

DNA sources: white onions, cauliflower, broccoli, garlic, split peas, bananas, and/or animal sources such as chicken liver, calf thymus

5   Knives   2   Blenders
         Cheesecloth or filter paper, 1-2   Microscopes
         Ice
1 L   SDS-NaCl-EDTA buffer solution (50 g sodium dodecyl sulfate, 50 g NaCl, 5 g sodium citrate, 0.2 mL of 0.5 EDTA and water to make 1 L)
500 mL Isopropanol (in an ice bath)

41 Digestion of Foods

A. Digestion of Carbohydrates

10   Spot plates (or plastic sheets)   Amylase preparation
200 mL 1% Starch   100 mL Iodine reagent
500 mL Benedict's reagent

B. Digestion of Fats

2–3   pH meters or pH paper   1 qt   Whole milk
200 mL 0.1 *M* NaOH   50 mL   Safflower oil
50 mL   Bile salts   100 mL 2% Pancreatin
100 mL 1% Phenolphthalein

C. Protein Digestion

2–3   Hard-boiled eggs   100 mL 2% Pepsin
100 mL 0.1 *M* HCl

42 Analysis of Urine

A. Color, pH, and Specific Gravity

3   Urinometers (hydrometers), *or* a small graduated cylinder and beaker
100 mL "Normal Urine" specimen   100 mL "Abnormal Urine" specimen

B. Electrolytes

10   Flame test wires   100 mL 3 *M* HCl
Place 100 mL in dropper bottles:

0.1 M AgNO$_3$ 0.1 M BaCl$_2$
3 M HNO$_3$ (NH$_4$)$_2$MoO$_4$ ammonium molybdate reagent

C. Glucose
20   Reagent strips: Clinistix, or multitest strips such as Multistix, Labstix, or Uristix
100 mL "Normal Urine"
100 mL "Abnormal Urine"
100 mL Benedict's reagent

D. Ketone Bodies
20   Reagent strips: Clinistix, or multitest strips such as Multistix, Labstix, or Uristix
100 mL "Normal Urine"
100 mL "Abnormal Urine"

E. Protein
20   Ketostix, Albustix, Labstix, or Multistix strips
100 mL 1 M HAc
100 mL "Normal Urine"
100 mL "Abnormal Urine"

F. Urobilinogen
20   Urobilistix, Labstix, or Multistix strips
100 mL "Normal Urine"
100 mL "Abnormal Urine"

Preparation of Solutions Used in the Laboratory

Acids and bases

Acetic acid C$_2$H$_3$O$_2$ (HAc)
    0.1 M HAc   Dilute 0.6 mL of glacial acetic acid with water to make 100 mL
    1 M HAc   Dilute 6 mL of glacial acetic acid with water to make 100 mL
    10% HAc   Dilute 50 mL of glacial HAc with water to make 500 mL

Ammonium hydroxide NH$_4$OH
    0.1 M NH$_4$OH   Dilute 6.7 mL conc. NH$_4$OH with water to make 1.0 L
    0.5 M NH$_4$OH   Dilute 34 mL conc. NH$_4$OH with water to make 1.0 L
    1 M NH$_4$OH   Dilute 67 mL conc. NH$_4$OH with water to make 1.0 L

Hydrochloric acid HCl
    0.1 M HCl   Dilute 8.3 mL of conc. HCl with water to 1 L and standardize against
                       standardized 0.1 M NaOH (or 0.2 M NaOH)
    1.0 HCl   Dilute 85 mL conc. HCl with water to make 1.0 L
    2.0 M HCl   Dilute 170 mL conc. HCl with water to make 1.0 L
    3.0 M HCl   Dilute 250 mL conc. HCl with water to make 1.0 L
    10% HCl   Dilute 230 mL con. HCl to make 1.0 L

Nitric acid HNO$_3$
    6 M HNO$_3$ Dilute 76 mL conc. HNO$_3$ with water to make 200 mL
    10% HNO$_3$ Dilute 10 mL conc. HNO$_3$ with water to make 100 mL

Sodium hydroxide NaOH
    0.1 M NaOH   Dissolve 4.0 g NaOH in water to make 1.0 L
                       Standardization: Weigh a 1-g sample of potassium hydrogen phthalate,
                       KC$_8$H$_5$O$_4$, to 0.001 g. Dissolve in 25 mL of water, add phenolphthalein

Nitric acid HNO_3

6 M HNO_3	Dilute 76 mL conc. HNO_3 with water to make 200 mL
10% HNO_3	Dilute 10 mL conc. HNO_3 with water to make 100 mL

Sodium hydroxide NaOH

0.1 M NaOH Dissolve 4.0 g NaOH in water to make 1.0 L

Standardization: Weigh a 1-g sample of potassium hydrogen phthalate, $KC_8H_5O_4$, to 0.001 g. Dissolve in 25 mL of water, add phenolphthalein indicator, and titrate with the prepared NaOH solution. Calculate the molarity (3 significant figures) as

$$\text{g phthalate} \times \frac{\text{1 mole phthalate}}{\text{204 g phthalate}} \times \frac{1}{\text{L NaOH used}} = \underline{\quad} M$$

6 M NaOH	Dissolve 240 g NaOH in water to make 1.0 L
3 M NaOH	Dissolve 120 g NaOH in water to make 1.0 L
10% NaOH	Dissolve 10 g NaOH in water to make 100 mL
20% NaOH	Dissolve 20 NaOH in water to make 100 mL

Salt solutions

Aluminum sulfate   1% $Al_2(SO_4)_3$ Dissolve 1.5 g $Al_2(SO_4)_3 \cdot 9H_2O$ in water to make 100 mL

Ammonium chloride   0.1 M NH_4Cl Dissolve 0.54 g NH_4Cl in water to make 100 mL

Ammonium molybdate   Dissolve 8.1 g H_2MoO_4 in 20 mL water. Add 6 mL conc. NH_4OH to give a saturated solution. Filter. Slowly add filtrate to a mixture of 27 mL conc. HNO_3 and 40 mL water. Let stand 1 day. Filter and add water to 100 mL.

Ammonium oxalate   0.1 M $(NH_4)_2C_2O_4$ Dissolve 1.4 g $(NH_4)_2C_2O_4 \cdot H_2O$ with water to 100 mL

Barium chloride   0.1 M $BaCl_2$ Dissolve 2.4 g $BaCl_2 \cdot 2H_2O$ in water to make 100 mL

Barium nitrate   0.1 M $Ba(NO_3)_2$ Dissolve 2.6 g $Ba(NO_3)_2$ in water to make 100 mL

Calcium chloride   1% $CaCl_2$ Dissolve 1.3 g $CaCl_2 \cdot 2H_2O$ in water to make 100 mL

20% $CaCl_2$ Dissolve 200 g $CaCl_2$ in water to make 1.0 L
0.1 M $CaCl_2$ Dissolve 1.5 g $CaCl_2 \cdot 2H_2O$ in water to make 100 mL

Copper(II) chloride   0.1 M $CuCl_2$ Dissolve 1.7 g $CuCl_2 \cdot 2H_2O$ in water to make 100 mL

Copper(II) sulfate
0.1 M $CuSO_4$	Dissolve 2.5 g $CuSO_4 \cdot 5H_2O$ in water to make 100 mL
1 M $CuSO_4$	Dissolve 25 g $CuSO_4 \cdot 5H_2O$ in water to make 100 mL
1% $CuSO_4$	Dissolve 0.78 g $CuSO_4 \cdot 5H_2O$ in water to make 50 mL
5% $CuSO_4$	Dissolve 15.6 g $CuSO_4 \cdot 5H_2O$ in water to give 200 mL

Iron(III) chloride
0.1 M $FeCl_3$	Dissolve 2.7 g $FeCl_3 \cdot 6H_2O$ in water to make 100 mL
1% $FeCl_3$	Dissolve 1.7 g $FeCl_3 \cdot 6H_2O$ in water to give 100 mL

Iron(III) nitrate
0.01 M $Fe(NO_3)_3$	Dissolve 0.40g $Fe(NO_3)_3 \cdot 9H_2O$ in water to make 100 mL
1 M $Fe(NO_3)_3$	Dissolve 40.g $Fe(NO_3)_3 \cdot 9H_2O$ in water to make 100 mL

Lead(II) nitrate)   1% $Pb(NO_3)_2$ Dissolve 0.5 g $Pb(NO_3)_2$ in water to make 50 mL

Magnesium chloride   1% $MgCl_2$ Dissolve 2.1 g $MgCl_2 \cdot 6H_2O$ in water to make 100 mL

Mercury(II) chloride   1% $HgCl_2$ Dissolve 0.5 g $HgCl_2$ in water to make 100 mL

411

Appendix

Potassium chloride	0.1 M KCl	Dissolve 0.75 g KCl in water to make 100 mL
Potassium permanganate	1% $KMnO_4$	Dissolve 1g $KMnO_4$ in water to give 100 mL
Potassium thiocyanate	0.01 M KSCN	Dissolve 0.097 g KSCN in water to make 100 mL
	0.1 M KSCN	Dissolve 0.97 g KSCN in water to make 100 mL
	1 M KSCN	Dissolve 9.7 g KSCN in water to make 100 mL
Silver nitrate	0.1 M $AgNO_3$	Dissolve 1.7 g $AgNO_3$ in water to make 100 mL
	1% $AgNO_3$	Dissolve 0.5 g $AgNO_3$ in water to make 50 mL
Sodium carbonate	0.1 M Na_2CO_3	Dissolve 2.9 g $Na_2CO_3 \cdot 7H_2O$ in water to make 100 mL
Sodium chloride	0.1 M NaCl	Dissolve 0.58 g NaCl in water to make 100 mL
	1% NaCl	Dissolve 1 g NaCl in water to make 100 mL
	10% NaCl	Dissolve 10 g NaCl in water to make 100 mL
	20% NaCl	Dissolve 200 g NaCl in water to make 1.0 L
	Saturated NaCl	Add 80 g NaCl to water to make 200 mL
Sodium phosphate	0.1 M Na_3PO_4	Dissolve 3.8 g $Na_3PO_4 \cdot 12H_2O$ in water to make 100 mL
Sodium sulfate	0.1 M Na_2SO_4	Dissolve 3.2 g $Na_2SO_4 \cdot 10H_2O$ in water to make 100 mL
	1% Na_2SO_4	Dissolve 2.3 g $Na_2SO_4 \cdot 10H_2O$ in water to make 100 mL
Strontium chloride	0.1 M $SrCl_2$	Dissolve 1.95 g $SrCl_2 \cdot 2H_2O$ in water to make 100 mL

Carbohydrates

Fructose	2% fructose	Add 1 g fructose to water to make 50 mL
Glucose	0.1 M glucose	Dissolve 1.8 g glucose in water to make 100 mL
	2% glucose	Add 1 g glucose to water to make 50 mL
	10% glucose	Dissolve 10 g glucose in water to make 100 mL
Lactose	2% lactose	Add 1 g lactose to water to make 50 mL
Sucrose	0.1 M sucrose	Dissolve 3.42 g sucrose in water to make 100 mL
	2% sucrose	Add 1 g sucrose to water to make 50 mL
Starch	1%	Make a paste of 2 g soluble starch and 40 mL water. Add to 160 mL of boiling water to make 200 mL. Stir and cool.
	2%	Make a paste of 4 g soluble starch and 40 mL water. Add to 160 mL of boiling water to make 200 mL. Stir and cool.

Reagents

Benedict's reagent	Dissolve 86 g sodium citrate, $Na_3C_6H_5O_7$, and 50 g anhydrous Na_2CO_3 in 400 mL water. Warm. Dissolve 8.6 g $CuSO_4 \cdot 5H_2O$ in 50 mL water. Add to sodium citrate solution, stir, and add water to make 500 mL solution.
DNA buffer	SDS-NaCl-EDTA buffer solution (50 g sodium dodecyl sulfate, 50 g NaCl, 5 g sodium citrate, 0.2 mL of 0.5 EDTA and water to make 1 L)
0.15% Salicylic acid	Dissolve 0.15 g salicylic acid in 100 mL water.
Seliwanoff's reagent	Dissolve 0.15 g resorcinol in 100 mL 6 M HCl
Iodine solution	Dissolve 10 g I_2 + 20 g KI in water to make 500 mL
Iodoform (iodine solution)	Dissolve 10 g I_2 + 20 g KI in water to make 100 mL
0.2% Ninhydrin	Dissolve 0.2 g in ethanol to make 100 mL
20% Phenol	Dissolve 20 g phenol in water to make 100 mL

4% Hexamethylenediamine/NaOH

       Combine 3.0 g $H_2N(CH_2)_6NH_2$ + 1.0 g NaOH and add water to 50 mL

4% Sebacoyl chloride/Hexane

       Combine 20.0 mL $ClCO(CH_2)_8COCl$ and hexane to make 50 mL

Indicators

1% Phenolphthalein	Dissolve 1 g phenolphthalein in 50 mL ethanol and 50 mL water
1% Bromine solution	Dilute 1 mL Br_2 with methylene chloride to make 100 mL
2% Chromate	Dissolve 2.0 g $K_2Cr_2O_7$ in 10 mL of 6 M H_2SO_4; then carefully add to water to make 100 mL

Amino acids

1% Alanine	Dissolve 0.5 g alanine in water to make 50 mL
1% Aspartic acid	Dissolve 0.5 g aspartic acid in water to make 50 mL
1% Glutamic acid	Dissolve 0.5 g glutamic acid in water to make 50 mL
1% Glycine	Dissolve 1 g glycine in water to make 100 mL
1% Lysine	Dissolve 0.5 g lysine in water to make 50 mL
1% Phenylalanine	Dissolve 0.5 g phenylalanine in water to make 50 mL
1% Serine	Dissolve 0.5 g serine in water to make 50 mL
1% Tyrosine	Dissolve 1 g tyrosine in water to make 100 mL

Proteins

1% Egg albumin	Dissolve 1 g egg albumin in water to make 100 mL
1% Gelatin	Dissolve 1 g gelatin in water to make 100 mL
2% Pancreatin	Dissolve 2.0 pancreatin in 100 mL 0.25% Na_2CO_3. Use immediately.
2% Pepsin	Dissolve 2% pepsin in water to make 100 mL

Urine specimens

"Normal Urine"	Mix 2 mL of 0.1 M NaCl, 2 mL of 0.1 M Na_2SO_4, 2 mL of 0.1 M Na_3PO_4, Add 1 g of urea. Adjust pH to 5.5–7.0 with 5 to 10 drops of 6 M HCl and dilute to 100 mL.
"Abnormal Urine"	To the recipe for "normal urine," add 1 g glucose, 1 g egg albumin, 4 mL acetone, and 1 mL 6 M HCl.